HOW WE EAT WITH OUR EYES

AND THINK WITH OUR STOMACHS

A . psychologist, **DIANA VON KOPP** has taught leadership c ɔetence, such as decision making and stress management, t line pilots for more than ten years. Fascinated by the fact t ood has an immense impact on our brain, performance, a vellbeing, she dove into research and has been writing for F *kfurter Allgemeine Zeitung*'s blog *Food Affair* ever since.

S e 2006, **MELANIE MÜHL** has been a features editor a *Frankfurter Allegemeine Zeitung*, Germany's leading r spaper, for which she also coauthors the *Food Affair* blog, ing hundreds of thousands of readers each month. She ed journalism at the University of Karlsruhe and at King's ersity College in Ontario, Canada.

in Germany, **CAROLIN SOMMER** spent a year in California a studied applied languages at universities in Germany, F ıce, and the UK. She now lives in the UK with her husband a three sons. Her previous translations include Jennifer ge's bestselling memoir, *My Grandfather Would Have Shot M*, published by The Experiment. For more information about C ɾolin Sommer, please visit www.carolinsommer.com.

HOW WE EAT WITH OUR EYES AND THINK WITH OUR STOMACHS

THE HIDDEN INFLUENCES THAT SHAPE YOUR EATING HABITS

MELANIE MÜHL & DIANA VON KOPP

Translated by Carolin Sommer

SCRIBE
Melbourne • London

Scribe Publications
2 John Street, Clerkenwell, London, WC1N 2ES, United Kingdom
18–20 Edward St, Brunswick, Victoria 3056, Australia

Published by Scribe 2017

Originally published in Germany as *Die Kunst des klugen Essens* 2016.
First published in the English language in North America by The Experiment, LLC, in 2017.
This edition published by arrangement with The Experiment, LLC.

Cover design by Sarah Smith
Text design by Sarah Schneider

Printed and bound in the UK by CPI Group (UK) Ltd, Croydon, CR0 4YY.

Scribe Publications is committed to the sustainable use of natural resources and the use of paper products made responsibly from those resources.

9781911617143 (UK paperback)
9781925322972 (Australian paperback)
9781925548846 (e-book)

CiP records for this title are available from the National Library of Australia and the British Library.

scribepublications.co.uk
scribepublications.com.au

CONTENTS

PART II. How You Diet with Your Brain

PART III. How You Savour with Your Ears

PART V. How You Feast with Your Feelings

PART VI. How You Choose with Your Tongue

Foreword

Although we eat a thousand meals every year, we're still not really sure why we ordered what we did for lunch. We might think we are master and commander of everything we eat, but the size of a plate, your friend next to you, and the fonts on the menu can trick you more than you could ever realise. But there's good news: While these circumstances can lead us to eat too much, they can also help us eat less—provided we make them work in our favour. Melanie Mühl and Diana von Kopp show you how.

How We Eat with Our Eyes and Think with Our Stomachs is the perfect page-turner for pointing out all of the hidden persuaders that cause us to eat too much of the wrong foods and to not enjoy food as much as we should. It's like an 'anti-diet' mystery novel. With no lecturing or finger wagging, Mühl and von Kopp perfectly balance great science-based findings with easy news-you-can-use tonight.

This book takes forty-two puzzling questions about why we eat the way we do and demystifies them by using clever examples and insightful science-based studies: Should you go to bed hungry? Why did your last diet fail? What soundtrack should you play during dinner tonight? Should you fight a chocolate craving or allow yourself a taste? These are more than just interesting conversation topics. The answers to these questions can change what we *eat* by changing what we *know*. You'll also learn about your food radius, the trophy kitchen syndrome, the doggie bag paradox, and what your love of spicy food says about your personality.

In the food world, some books are often old wine in a new bottle. This book is a serious departure. It allows you to pick and choose the topics you find most interesting and relevant to yourself in the same way you would if you were in a dinner conversation. You can use this for pure enjoyment, just like the dinner conversation, or to really tune in to your senses and even to become a better person. You'll leave the table knowing lots of tricks that will make you smarter and healthier the next time you sit down to eat.

This is the future of healthier eating. It shows how we can make simple changes in our environment to change what we eat. We don't need to use willpower or 24/7 mindfulness. We just need to know more about what Mühl and von Kopp already do—and put it into action.

Of all of the books I've read on food, psychology, and eating behaviour this year, *How We Eat with*

Our Eyes and Think with Our Stomachs is far and away the most interesting, useful, and entertaining.

Brian Wansink, PhD
Author of *Mindless Eating* and *Slim by Design*
Director of Cornell University's Food and Brand Lab

Introduction

FIRST THE GOOD NEWS: We live in a veritable food paradise that allows us to satisfy our cravings around the clock. The Garden of Eden, by comparison, was a joke. At the same time—and this is where the not-so-good news comes in—food is becoming increasingly complicated. Somewhere between vegetarianism, the paleo diet, the raw-food diet, the blood-type diet, and juice cleanses, we've lost track of how, when, and what we're supposed to eat and how to have a carefree relationship with food—despite the fact that eating is one of the most sensual experiences of all!

In order to re-create that carefree relationship, and to understand why we act (and eat!) the way we do, we must peek behind the curtain—that is, behind our own decisions. The number of food choices we make each day is surprising: over two hundred. It goes without saying that not each and every one of those choices is a conscious one, and we obviously don't always analyse in the moment why we ordered dessert or why we sometimes prefer pepperoni pizza; quite the opposite. This is where our subconscious comes in, doing the job for us. On the one hand, it's useful as it frees up brain capacity for other things, but on the other hand, it's dangerous to hand over the reins to a part of our brain that we're not aware of—and that's exactly what we do with the highly sensitive matter of food. But how are we to find our way through this food jungle, eat more healthily and wisely, and really enjoy our food when we can't even say exactly why some foods make us happy while others make our stomachs turn? Or why we sometimes eat so much that we fear we might burst? How do we know when we've had enough? And by the way, how does taste work, and what are the roles of our psyche and our brain in our daily food choices?

Forget for a minute the many diet myths that we're constantly being fed, which only make us insecure. Instead, let's focus on the verified findings from behavioural psychology and neuroscience that decode our sense of taste and shine a light on the social aspect of food. In recent years, rapid progress has been made in research into these areas, unearthing some astonishing facts—a stroke of luck not just for our health, but also for our gustatory pleasure.

And how to understand our food preferences? These foundations are already laid in the womb. Did you know: the sweeter the amniotic fluid, the more often the unborn baby swallows? Bitter compounds, however, are not so popular. Once we're born, the conditioning continues. Some of us turn into picky eaters, while others happily eat everything that's put in front of us. Sooner or later we find ourselves on our first diet and realise: Darn, it's not working! But why not? Because, in short, we are not rational eaters.

Pointing out the irrational way we often make choices, the behavioural psychologist Dan Ariely describes us humans as 'pawns in a game whose forces we largely fail to comprehend'. And when we do, we systematically underestimate them. This is also true with food. The aim of this book is to expose these forces and, using this information, to help improve our daily lives. Let's take back the reins and start eating with more awareness and more enjoyment!

PART I

How You Eat with Your Eyes

CHAPTER 1

The Colour of Flavour

Is there such a thing as blind faith at the dinner table?

In 1964, Alfred Hitchcock, famous for his morbid sense of humor, invited a few friends to a Christmas party. Among them were Cary Grant and his then wife Dyan Cannon. 'Be prepared for *anything*,' Grant warned her as they pulled into the drive of the Hitchcocks' Bel Air home. Hitchcock wasn't known as 'the master of the unexpected' for nothing: The film director greeted them with a tray of Windex-blue martinis and said, 'I hope you'll forgive me, Cary, but we're fresh out of LSD. I hope a martini will suffice. I made them so you could have a drink and see colours at the same time.' That's how Cannon remembers the scene in her memoir *Dear Cary: My Life with Cary Grant*. Later, when she and Grant had taken their seats, Cannon writes:

> Two butlers brought large, covered plates to the table. Hitch gave them a nod, and they removed the covers to reveal large slabs of prime rib. The beef smelled wonderful, but it looked awful. It was blue. Bright, turquoise blue. Then along came the side dishes: blue broccoli, blue potatoes, blue rolls. . . . 'Do you think it's safe to eat?' I whispered to Cary. 'The color may be off-putting, but I'm sure it's perfectly fine,' Cary said sanguinely. He was wrong. By the time the night was over, the two of us had worn a groove in the carpet between the bed and the bathroom.

Whether it was the meat that was off or its blue colour that was ultimately to blame for their upset stomachs, we shall never know. But one thing is certain: We instinctively recoil from food that has the 'wrong' colour. Because vision is our dominant sensory input, a change in a food's colour can overwhelm our other senses and lead to a false positive for taste, so to speak. This phenomenon was first demonstrated in a scientific study conducted by the chemist H. C. Moir in 1936. In the study, Moir gave his colleagues jelly in various colours—with no difference in their flavour—and noticed that they each reported tasting the flavour associated with the colour that they saw. They trusted their eyes—and the associations triggered by the colours—more than their taste buds. According to

Kathrin Ohla, a psychologist at the German Institute of Human Nutrition at Potsdam-Rehbrücke near Berlin, optics play a particularly important role in our perception of flavour. She explains, 'It's because the mere taste sensation is not object-related enough. When I taste something sweet or sour, that on its own doesn't tell me what it is.'

Twentieth-century psychologist Karl Duncker also studied this colour phenomenon. In 1939 he conducted a study that tested and assessed a new product on the US market: white chocolate. Volunteers who were allowed to see the chocolate while eating it described it as milkier and less flavoursome than dark chocolate, whereas volunteers who did not see it beforehand did not report noticeable differences. Since then, numerous studies have shown how visual cues shape and manipulate the way we perceive flavour. In the late eighties, subjects in a test were given vanilla custard with the colour of chocolate to eat. None of the participants recognised the vanilla flavour. In some of the most famous experiments, even renowned wine experts failed to identify the distinct taste of white wine when it had been dyed red.

Colour experiments under market conditions are extremely risky, even when they are supported by elaborate advertising campaigns. Coca-Cola had that painful experience in 1993. One year after Pepsi made a similar attempt with a product called Crystal Pepsi, Coca-Cola launched their own sugar-free, clear cola called Tab Clear in the United States, the United Kingdom, and Australia. Customers reacted to the missing characteristic colour by rejecting the product; Tab Clear proved to be a spectacular flop. Consumers didn't recognise *their* Coke. Coca-Cola had taken sensory incongruence a step too far, separating their product from the nostalgia—and colour—associated with it. Twelve months later Tab Clear was pulled off the shelves.

It's a well-known fact that the food industry uses colours such as synthetic beta-carotene (an orange-yellow dye) in an attempt to manipulate customer behaviour. Take margarine, for example: Its natural colour is really more of a white, and its taste is oilier than that of yellow butter. The addition of beta-carotene makes margarine look more like butter, and it appears creamier than it really is. The 'margarine question' goes back surprisingly far. In 1895, C. Petersen gave a lecture with that title at the general meeting of the Association of the German Dairy Industry in Berlin, which included this comment about the colour of margarine: 'We'll have to raise the question as to why margarine is dyed the colour of butter, and the only possible answer to that question is: because it is believed that it will make people think that they are in fact consuming butter.' And even if this addition of colour was presented as harmless, he added, it was still done 'for the purpose of deception'. In modern consumer psychology, such deception goes by the name of 'product experience'.

Playing with sensory incongruence is a popular game not only among conglomerates like Coca-Cola but among celebrity chefs as well. And they, too, walk a fine line. If it goes well, the customer is thrilled and remembers the event as an extraordinary taste experience. If it goes wrong, however, the customer won't be back.

FOOD FOR THOUGHT: We appraise our food through vision first and do not approach the table blindly. It may be fun to trick your dinner guests, but draw the line at blue beef à la Hitchcock. Even the most adventurous gastronomes should think twice when colour doesn't look quite right—or is too convincing.

Consider the many ways in which artificial food colouring sneaks into your diet—and what its purpose is there. Do you enjoy minty-green ice cream? What about fruity drinks or blue martinis? What are your eyes and your brain suggesting to you about what a food should taste like before you actually eat it?

CHAPTER 2

A Plate of Art

How much does it cost to eat with your eyes?

WASSILY KANDINSKY'S *PAINTING NO. 201*, PART of a famous quartet, can be admired at the Museum of Modern Art in New York. Depending on one's mood and needs, it is possible to see all kinds of things in this abstract, colour-intensive painting. One art critic saw the four seasons in the series, while others hold that Kandinsky had a landscape in mind. Charles Michel, the chef in residence at the Oxford Crossmodal Research Lab, saw something entirely different: He spotted a salad, including a prominently positioned mushroom in the upper left-hand corner. The 'Kandinsky salad' was born. Inspired by his own food art, Michel went on to examine how our expectations, our taste, and, above all, our willingness to spend money can be influenced by how our food is arranged on the plate.

For the experiment, Michel and his team divided their volunteers into three groups. Each group was served the same salad—only the presentation varied. For the first group, the components of the salad (including mushrooms, broccoli, and snow peas or mange tout) were placed in neat rows on the plate. The second group was served theirs as a simple tossed salad of healthy ingredients, and the third was presented with an artistically arranged salad. They had no idea that their salad had been arranged to look like the Kandinsky painting. Before and after eating the salad, participants were asked to rate its presentation, how tasty it looked, and how much they'd be willing to pay for it. The result: The 'Kandinsky salad' won hands down. The participants gave the Kandinsky plating the highest ratings in all three categories and were even willing to pay more for it—before and after they tasted it. 'Diners intuitively attribute an artistic value to the food, find it more complex and like it more when the culinary elements are arranged to look like an abstract-art painting,' Michel concluded.

'You eat with your eyes first'—a truism that it seems can't be repeated often enough. In any case, chefs at top restaurants around the world appear to be engaged in some kind of unofficial competition in creativity and finesse. The best chefs are creating real—but very temporary—works of art. One of the greatest destinations is Noma in Copenhagen, where a multicourse dinner will set you back several hundred euros. Some of their dishes would not look out of place in a museum themselves, such as the mossy landscape served in a clay bowl with crispy lichen that look like sponges and are to be dipped in crème fraîche.

Whether cooking is an art (and if so, to what extent) is up for debate—and a different subject altogether. There is a long tradition of top chefs using artistic presentations of their dishes to mesmerise customers until they are too stunned to know if they should look at their food or eat it. Today's trends favour minimalism and clarity, but it wasn't always like that. Published in Nuremberg in 1642, the *Trincir-Buch*, a book that deals with, among other things, the carving and presenting of food, describes dishes called *Schauessen*: 'food for show—that are made by hand, lovely to look at, and can be enjoyed as well. They please the eyes first and then the mouth, and they are usually brought out once everyone has satisfied their appetite with other dishes.' These showy dishes, culinary arrangements associated with the nobility, were towers of decadence that primarily served as a status symbol. After all, not everything in those displays that looked like food was edible.

And while the displays spread on dining tables back when Blue Bottle operated a café at the San Francisco Museum of Modern Art (SFMOMA) were edible, they looked almost too beautiful to eat. The creator behind this culinary art, the pastry chef Caitlin Freeman, nearly became a photographer instead. The turning point arrived when she happened across the 1963 painting *Cakes* by the American pop artist Wayne Thiebaud, and became obsessed with cakes. As the pastry chef at Blue Bottle at SFMOMA, she created cakes inspired by artworks. Her masterpiece was a 'Mondrian Cake', an edible canvas of red, white, blue, and yellow sponge cake divided into geometric blocks and encased in chocolate icing. Baking this piece of art at home is quite a time-consuming affair. In her book *Modern Art Desserts*, Freeman recommends allowing six hours of intensive work over two days, but that's still no guarantee that the result will end up looking like an actual Mondrian Cake and not like a food-colouring experiment gone wrong. For significantly quicker results, the book offers recipes inspired by Warhol, Ryman, Matisse, and Lichtenstein—all fashioned into cake.

FOOD FOR THOUGHT: It's no surprise that we react not only to how a food tastes but also to how it looks. Use it to your advantage! The next time you want to impress your guests, take a few extra minutes and tap into your creative side when plating. Think about which colours look best near each other, and which might clash. Some quick tips: Set the table. Try varying your dishes with different-shaped plates. Wipe off dripping sauces (neat is always better!). Vary colour, texture, and height—the plate doesn't all have to be flat; try stacking different foods on the plate. Apply garnishes—they not only add intrigue to the plate but add a layer to the bite!

CHAPTER 3

Dish Decisions

How does plating affect your appetite?

The weekend is approaching and you have invited a few close friends over for dinner. One of them will bring her boyfriend, a know-it-all without a sense of humor to speak of. Being the exemplary host that you are, you want all of them to enjoy their food, of course—but maybe not to the same degree. So you start thinking about how you can manipulate the taste experience of your guests. All you need is a little time to prepare. Check out your china cupboard, and do it early—don't wait until the big day. What do you see? Big plates? Small plates? White plates? Square plates in different colours? You likely have the usual white plates, and that is precisely the problem.

If you want to prevent the humorless know-it-all from helping himself to one slice after another of your prime roast beef, don't serve it to him on a white plate; give him a red one. Scientists at Oxford University discovered in neurogastronomic experiments that red tableware reduces hunger. We associate the colour red with danger: Toadstools are red, as are warning signs, and fire extinguishers. Our natural response to danger is to run away, not to feel hungry. Even sushi fans have to admit that six salmon nigiri don't look very enticing on a red plate. As a result, we tend to eat less because we enjoy the food less, which adds a whole new dimension to the saying 'you eat with your eyes first'.

Tableware is a serious matter: A plate is like a frame for the food presented on it. The Japanese have known this for a long time. They were already agonising over the best possible presentation of food when the French and the rest of the world were still slopping theirs mindlessly onto their plates. No Japanese restaurant will ever make you feel that the plates are too small for the food they hold—on the contrary. By now, an artistically inspired science of plating has developed that seeks to transform appetising dishes into alluring ones. It wants us to love our food before we've even picked up our knives and forks. Or it might want the opposite, which brings us back to our unwelcome guest.

This may be a bitter pill to swallow for those who believe that the quality of the ingredients and masterful preparation alone are the key factors for a successful meal. Even the lovingly cared-for Japanese Kobe beef, which can sell for as much as three hundred US dollars per pound, won't taste quite as sensational if presented on a red plate. Food commandeers all of our senses. Not only do we see, smell, and taste food, but we also hear it as we eat and feel it on our tongues. We enjoy it on a multisensory level.

Dessert shouldn't be carelessly served on any old plate either. The Spanish celebrity chef and former head chef at elBulli, Ferran Adrià, led a study in which the participants were served a strawberry mousse either on white or on black plates. The mousse served on white plates fared much better in overall ratings: The testers rated its flavour 15 per cent more intense and 10 per cent sweeter than the same mousse served on black plates.

But why would that be? One reason was the contrasting colours. Strawberry mousse on white looks more enticing than on black, but colour associations also come into play. We look at a plate and within fractions of a second, we have formed our expectations—positive as well as negative ones. We usually try those foods first that look most appealing to us. That could be the steak, the sweet potato, or the beans.

Another example: We know that food in hospitals and nursing homes is rarely sublime and often atrocious. It smells dull, it looks dull, and it tastes dull. At its worst, it stops patients from eating altogether. Another experiment by scientists from Oxford University showed that dementia patients in a British hospital ate nearly a third more of their food if their white fish was served on blue plates instead of the usual beige ones: The fish was no longer an indefinable goop but now looked like it was freshly caught from the sea. A simple trick with great effect.

FOOD FOR THOUGHT: It's handy to have plates in a variety of shapes and colours. The day will come when you will be glad to have a red plate on hand. If you'd like a subtle reminder not to eat quite so much, try reaching for a salad plate rather than a dinner plate. You may even lose a few kilos that have been sticking around.

If you aren't confident in a dish you've cooked for guests, try playing up contrasts on the plate. Are there other ways to manipulate the colours on the plate in your favour? Simple visual tricks go a long way.

CHAPTER 4

All You Can ~~Eat~~ See

How do you control how much you eat at the buffet?

THE OWNER OF A LARGE CHAIN of Chinese all-you-can-eat buffets in the Midwest was in despair over his customers' appetites. Once they started eating, they would never stop. Magically attracted by the buffet, they piled the food high on their plates and devoured huge amounts. Apparently they interpreted the all-you-can-eat offer as a challenge. Driven by instinct, humans eat like they did in the Stone Age: stocking up for leaner times. The buffet chain's profits were hit hard by this gluttony. It had to stop.

And where does all the food go? It's hard to imagine that scarfing down many meals' worth of food doesn't cause extraordinary pain as the stomach fills up, yet we continue to eat. A study in the *Journal of Neuroscience* found that our tendency to feel pain is reduced when tasty food is on hand. It's paradoxical:

We seek pleasure and avoid pain, but we endure pain to attain pleasure. Nowhere is this more apparent than with food. When presented with chocolate, for example, we keep eating even after we've had our fill because it induces pleasure, and we don't want that sensation to end.

But how do you stop people from storming the all-you-can-eat buffet and encourage them instead to show restraint and eat in moderation? The Chinese-restaurant owner approached Brian Wansink, the director of the Food and Brand Lab at Cornell University, with just this question. Wansink and his team took up the challenge. They started by observing the customers' behaviour from the moment they entered the restaurant: how and where they went, where they sat, what they did with their jackets and with their napkins. Every detail counted, including the patrons' waistlines. How many of them were slim? How many overweight? How many obese? And how was it possible that some all-you-can-eaters stayed slim while others piled on kilo after kilo? What do we not yet know about the psychology of feasting?

The observers soon realised that there were stark behavioural differences among the individual diners. While slimmer people would often reach for small plates and chopsticks, larger customers would more likely pick up large plates and cutlery. So far, so good—no surprises there. But they also noticed that slimmer customers sat with their backs to the buffet, while the larger ones sat facing it. As a result, no diner replenishing their plate escaped notice, and every visitor to the buffet triggered the impulse to do the same—for fear of missing out—resulting in more returns to the buffet among customers who sat facing it. While diners with a low body mass index (BMI) inspected the food, 'explored' the buffet, and circled it at least once, those with a higher BMI made a beeline for the presented food and helped themselves seemingly at random. At the table, they chewed each mouthful eleven to twelve times, while the slimmer ones chomped around fourteen times. You have to ask yourself, how close did Wansink's spies come to the individual diners in order to make these observations?

Wansink's advice to the restaurant owner: Stock up on chopsticks. Instead of a selection of different-sized plates, provide mostly small plates and ensure that all diners are seated as far away from the buffet as possible. Better still, ensure that they can't see it at all from their table. In terms of layout this means erect screens to block the view of the food. If, out of the corners of their eyes, people can see enticing dishes and smell their delicious aromas just a few steps away, they will continue eating, whether they are full or not. Food that is available with minimal effort will be eaten.

Of course, there is a long tradition of mindless food consumption; culinary orgies were popular even in ancient Rome. The philosopher Seneca wrote about the 'voracious maws' of gluttons who were so lazy that they wished all they had to do was open their mouths like young chicks open their beaks. At the table, they would sit with their sluggish bodies slumped, lifting their hands less and less, leaving all the work to their cooks and servants. According to the book *Das römische Gastmahl: Eine Kulturgeschichte* (*The Roman Feast: A Cultural History*), attendant slaves would be tasked with 'deboning the fish, opening scallops, and removing bones from the meat—a new fashion that prompted Seneca to predict sarcastically that in the future, food would be served pre-chewed since the cooks appeared to be already doing the job of the teeth'. A very popular indulgence in Emperor Nero's day was 'a dish of clams, shelled oysters, deboned goatfish, birds, and sea urchins—a mix which was then covered in one and the same sauce'.

FOOD FOR THOUGHT: Eating with neither rhyme nor reason runs through history like a red ribbon, which takes us back to our gluttonous eaters in the Midwest. The owner of the Chinese restaurant chain suddenly saw his buffets with fresh eyes. Rather than rethinking his food options, he heeded Wansink's advice. This allegedly yielded an annual savings of thirty-eight thousand dollars in each of his restaurants.

CHAPTER 5

Supermarket Schemes

Why do you always fail to stick to the shopping list?

OUR TRUSTED SUPERMARKET IS ONE OF the most frequently visited places in our personal food radius (see 'The Food Radius', page 148). Every time we step through its doors we enter into a battle with our enemy, the supermarket corporation, whose motives are totally opposed to our own. While we hope to do our shopping quickly and efficiently, buying just the milk, bread, yoghurt, pesto, apples, and leeks on our list, the supermarket owners would prefer that we spend hours and hours strolling slowly through the aisles and reaching for new products left and right; products that we neither need nor want, which is why supermarkets use subtle psychological tricks to entice us. That's nothing new. We think we can see through their schemes and are convinced that we won't be conned. But we're wrong. Just because we understand something on a rational level does not make us immune to it.

It's only logical that supermarkets around the world follow certain layout principles and that we instantly know our way around them wherever we are. You will never find toilet paper and tissues right by the entrance. First impressions count, and in a supermarket the primary goal is to convey freshness, which is achieved by the scent of freshly baked goods and prominently displayed juicy fruit and crisp vegetables. In some supermarkets, vegetables are sprinkled with tiny droplets of water to give them a shiny, fresher appearance, but in reality this only makes them rot more quickly. What is true for vegetables applies to fruit as well. Bananas, for instance, have gone through a colour-optimisation process when planted. 'Each color represents the sales potential for the banana in question,' the neuromarketing expert Martin Lindstrom writes in his book *Brandwashed: Tricks Companies Use to Manipulate Our Minds and Persuade Us to Buy*. 'For example, sales records show that bananas with Pantone color 13-0858 (otherwise known as Vibrant Yellow) are less likely to sell than bananas with Pantone color 12-0752 (also called Buttercup), which is one grade warmer, visually, and seems to imply a riper, fresher fruit.'

The reason why the chilled aisles are located at the back of the shop is that dairy products are among the most frequently bought items (technically known as 'fast-moving consumer goods'). And so the consumer is made to walk past numerous other products—temptations. Expensive, branded products are placed at eye level (at a height of between 140 and 180 centimetres); cheaper ones (flour, sugar, etc.) are on the top and bottom shelves.

Often price isn't the decisive factor. In an interview, Lindstrom suggested, 'We tend to prefer brands that we have an emotional connection to. I call this a somatic marker. It's a bit like a bookmark in an Internet browser that keeps taking you back to the same website. Our brain has very similar bookmarks that make us reach for the same brands, which we do because we link them back to happy memories from our childhood or to other positive associations.' The scent of Johnson's baby powder, he claims, is one of the world's favourite smells.

If you've never looked closely at the floor of your local grocery shop, you should do it: Its structure affects your walking speed. Pushing your shopping cart over a grooved floor makes you feel that you are walking faster than you really are—and you automatically slow down. Slow music makes you move more slowly still and increases the likelihood of spontaneous purchases.

It's quite obvious why supermarkets sometimes set up little stands offering morsels of cheese, salami, or other interesting foods: They are actively promoting these products. What is not so obvious is that the idea behind this sale-boosting technique is the phenomenon of reciprocity. Reciprocity is an 'ancient technique', Rolf Dobelli writes in his bestselling book *The Art of Thinking Clearly*—it works along the lines of 'you scratch my back and I'll scratch yours'. He explains that psychologist Robert Cialdini, who studied the phenomenon, established that 'people have extreme difficulty being in another person's debt'. In other words, there's no such thing as a free lunch; the friendly snack is 'gentle blackmail' in disguise.

In his 1957 *New York Times* bestseller *The Hidden Persuaders*, journalist and consumerism critic Vance Packard quoted noted marketing strategist Louis Cheskin, who wrote that marketing researchers use 'techniques designed to reach the unconscious or subconscious mind'. These techniques are no longer just reaching for the subconscious mind; they are now holding it in their clutches.

According to Martin Lindstrom, visuals dominate the marketing world for now, but the future belongs to a multisensory approach to marketing: 'A British supermarket carried out an experiment whereby they played different types of music in the wine aisle. During the first week, they played accordion music, and oompah band music during the second week.

And even though the experiment was based on national clichés, it had an immediate impact on the sales figures: The first week saw a marked increase in the sale of French wines, the second week saw greater sales of German wines.'

FOOD FOR THOUGHT: Do you often run to the supermarket for one or two things only to emerge half an hour later laden with groceries? It isn't entirely your fault—but it is preventable! Know exactly what it is that you need to pick up and *stick to the shopping list*. You'll save yourself time, space in the pantry, and bucketloads of money. In the fight against the manipulation machine that is the supermarket, a helpful defence is tunnel vision and grim determination—if that doesn't seem to work, try putting on headphones and listening to music to keep you focused while you shop!

CHAPTER 6

Celebrity-Advice Fairy Tales

What qualifies celebrities to give health advice—and why do you listen?

LET'S DO A LITTLE THOUGHT EXPERIMENT: Imagine you are determined to change your eating habits and to eat more healthily. You plan to eat more fruit and definitely more vegetables. You go to your local book shop to look for books to support you on that journey and you bump into an acquaintance. You start talking. She is overweight and is looking for similar reading material. Diet is a big issue for her, too, but unlike you she has already read numerous books on the subject, attended seminars, tried recipes, and consulted a dietitian—and now she's giving you her unsolicited advice. How do you react? You're unlikely to be euphoric. No, judging by your acquaintance's waist, she's the last person you'd ask for competent diet advice. Well, you might be wrong. But where *are* you getting diet advice from?

Back home on your sofa, you're flipping through the pages of a lifestyle magazine and find the former supermodel Elle Macpherson looking up at you. Today, Elle Macpherson is engaged in the antiageing market. She is fifty-three, but could easily pass for forty-three. Her body? Amazing. Her message? Simple: Feel good, feed your cells, and you'll look great. Her own personal 'superfood' is neither goji berries nor chia seeds but an alkalising food supplement that she developed: 'The Super Elixir'. It's a green miracle powder that is said to promote vitality and wellbeing, as well as reduce stress, tiredness, and premature ageing of the skin. Price: $135 per box. You seriously consider ordering the powder.

Oscar-winning actress Gwyneth Paltrow also likes to present herself as a nutrition guru and posts tips on her website *Goop* that are not unlike the ones you see in women's magazines—the crucial difference is that her celebrity status affords her an authority and a credibility that moves her into the rank of an expert, a position for which she doesn't qualify. Just as you judged your acquaintance in the book shop, we judge celebrities by their success and good looks and assume that they possess intelligence and competence. If only *we* looked like that! And were as successful! Therefore, there must be some truth in their recommendations. Right?

We've fallen into a trap—the 'halo effect', a well-documented cognitive bias. The shine from a perceived saint's (or celebrity's) halo can make us blind to what's real. We project success in one area onto other areas, concluding that a person has expertise in an area that they do not. Objective evaluation criteria? Who needs them! The effect is particularly strong on those who are impulsive: The more impulsive a person is, the quicker they are to judge, and the greater the chance of falling into the halo effect trap. The halo effect is nothing new; the concept was first introduced nearly a hundred years ago by the American psychologist Edward Lee Thorndike. During the First World War, Thorndike researched how command personnel assessed their subordinate soldiers. He asked officers for an assessment of the soldiers' physique, character, leadership qualities, and intellect. Soldiers with an attractive face and ideal posture were consistently rated higher than their less physically endearing colleagues. Since then extensive research into cognitive distortion has been carried out, and such distortion has been demonstrated in many areas of our lives.

The influence celebrities can exert on our eating habits is augmented by another trend: namely, that food today is no longer just nourishment but has become a lifestyle statement. One of the magic words customers fall for hook, line, and sinker, and one that Paltrow & co. repeat over and over, is *detox*. We are urged time and again to cleanse our apparently permanently clogged insides. Everywhere we turn there is talk of how we need to cleanse the body. Humans are sinners: We eat sugar and wheat, drink coffee and alcohol, and binge on burgers and fries. Our miserable habits acidify our bodies—especially on weekends and holidays. Detox is the penance. But the truth is: A healthy body doesn't need to detox! It has a near-perfect, organic cleansing system that conveniently works day and night. Should the body actually be poisoned, a green smoothie is unlikely to help.

FOOD FOR THOUGHT: So, truthfully, whose advice are you more likely to take: that of your well-read acquaintance or that of your favourite health-centric celebrity? What other advice are we encouraged to take from unqualified sources who enjoy the illusion of authority? When confronted by one of those relentless marketing campaigns launched for nutrition products with a celebrity at the helm, ask yourself if that celebrity's image engenders trust—and authority. Are you being sold a product or a lifestyle?

The next time someone gives you some well-meaning advice, don't be too quick to judge. Sit back and verify the competence of the advisor before you take—or leave—it.

PART II

How You Diet with Your Brain

CHAPTER 7

The Raw-Food Fallacy

How does cooking make you smart?

Chimpanzees prefer cooked sweet potatoes to raw ones, just like humans. At first glance this is hardly surprising; after all, we share 98.5 per cent of our genes with these apes. This raises the question: Why don't chimpanzees spend more time in the kitchen making omelets? Harvard anthropologist Richard Wrangham thinks he has the answer: It's the cultural technology of cooking itself that has made us human. Without such skills as roasting and frying we would not be walking upright and might still be living in trees. In his book *Catching Fire: How Cooking Made Us Human*, Wrangham writes that cooking is basically in our genes. Humans have adapted to eating cooked food in the same way as cows have adapted to eating grass, he explains. This may sound strangely archaic in an age of microwaves and mass food production, an age in which cooking has become a televised event best enjoyed from the sofa, where we watch experts snip and chop. After all, cooking these days can mean as little as shoving a pizza in the oven. Yet, it was the discovery of fire and the preparation of cooked meals that laid the foundation for our intellectual development in the first place—and that includes the writing of scripts for cooking shows.

For our ancestors, the transition from raw to cooked food was a boon for efficiency. With a diet of tough, fibrous plants, chewing alone took about six hours each day, and digestion required a lot of energy. Cooking made our food more tender, which meant it spent less time in our stomachs, and we were able to extract more nutrients from it. Because food was more easily digestible, our digestive organs could shrink, while the extra energy obtained from cooked food allowed our brains to grow. Cooking made humans intelligent!

A study at the University of Parma revealed that by nibbling raw carrots we absorb only 1 per cent of their beta-carotene (good for eye health!), compared to 30 per cent when we eat them cooked. Mediterranean herbs such as thyme, rosemary, and sage require heat in order to develop their cell-protecting properties. The antioxidant potency of carrots, broccoli, and zucchini (courgette) increases when they are heat-treated. Study leader Nicoletta Pellegrini's conclusion confirms Wrangham's 'human cookivore' theory: 'Optimized processing increases the nutritional value of vegetables.'

Cooking itself has significant advantages, since it destroys toxins, kills pathogens, and prolongs shelf life. Once this was discovered, meat no longer needed to be consumed immediately but could be stored for longer periods of time. 'The adoption of cooking must have radically changed the way our ancestors ate, in ways that would have changed our social behavior too,' Wrangham writes. Instead of eating on their own, like chimpanzees, early humans assembled for a meal around the fire. They celebrated their hunting successes and shared their prey with those who returned empty-handed. They developed new preparation tools and techniques and fashioned knives from flints, pans from slabs of rock, and pots from turtle shells. The women collected and prepared the food, while the men went out to hunt. This sexual division of labour could only work because their food was cooked, Wrangham believes: 'A man who has spent most of the day hunting can satisfy his hunger easily when he returns to camp, because his evening meal is cooked. But if the food waiting for him in camp had all been raw, he would have had a major problem.' Because he would have had hours of chewing to do before he could get some well-deserved rest.

Such complex social changes required an enormous adaptive effort on the part of the brain. Neuroscientist Gordon M. Shepherd suspects that the rich aromas of cooked and fermented food kicked off the development of secondary, retronasal olfaction—that is, the scents we experience while eating—and thus the 'unique human brain flavor system'. After all, the aromas of food affect many processes of our neurons: emotions, memory, language, the control of our senses and appetite.

The invention of cooking is therefore one of the most successful chapters of human history. By increasingly and frivolously handing this cultural achievement over to industry, we are also losing a piece of our history, to say nothing of the high levels of hydrogenated fats, refined sugar, salt, and nutrient-poor white flour to which we subject our bodies. We are getting sicker and fatter, just like our pets, by the way, whom we're feeding industrially manufactured food. It is plain to see that cooking with fresh and natural ingredients is best for our health.

FOOD FOR THOUGHT: The art of cooking is in crisis! The fact that so many of us shun the stove is ironic given how much and how passionately we like to talk—and argue—about food and diet. But 'eating smart' does not just mean talking about food, it also means cooking food. When was the last time you cooked a meal from scratch? If it's been a while, you may want to get back in the kitchen for a complete mental workout: You need strategy, coordination, and memory; not to mention cooking engages all of your senses.

CHAPTER 8

Gluten Anxiety

Why do you keep falling for food myths?

A REGULAR FEATURE ON JIMMY KIMMEL'S late-night talk show is 'The Pedestrian Question', wherein a camera crew goes out on the street to interview people. For one of his shows, Kimmel asked people exercising in a park if they maintained a gluten-free diet. They replied, 'I do,' as if it were the most natural thing in the world. The next question, 'What is gluten?', however, left them nonplussed. None of them had a clear answer and speculated wildly: Is it a grain? Pasta, pizza, bread? But all the interviewees agreed on one thing: Gluten is bad. Someone had recommended they stop eating it—the yoga teacher, the fitness coach, or the girlfriend who had just read about it in a book.

Still not sure what gluten is? In case you didn't know, gluten is a mixture of proteins. Full stop. Those proteins come from two different groups: the prolamines and the glutelins. In wheat, these are gliadin and glutenin, but they're different in other grains like rye or barley. When the two proteins get wet and combine, they form gluten.

To many, not eating gluten says, 'I take care of my body and am happy to make sacrifices for it by avoiding foods like pizza, cake, and bread.' But sacrifice is driven by more than just mindfulness; books such as William Davis' *Wheat Belly: Lose the Wheat, Lose the Weight, and Find Your Path Back to Health* and David Perlmutter's *Grain Brain: The Surprising Truth About Wheat, Carbs, and Sugar—Your Brain's Silent Killers* can stoke real fear. Dr Perlmutter warns that wheat has a hold over us in the same way that heroin has a hold over a desperate addict. 'Gluten is what I call a "silent germ." It can inflict lasting damage without your knowing it.'

Put in such exaggerated terms, this apparent threat seems akin to the medieval doomsday scenario of Saint Anthony's fire (also known as ergotism): Certain grains become infected with a fungus—the ergot—causing a usually fatal condition. Anyone who eats grain products contaminated with the fungus is at risk. Is gluten just as lethal? It negatively affects only one in a hundred people: people who suffer from coeliac disease, a condition in the small intestine that can be treated with a gluten-free diet.

In America, the nonprofit Gluten-Free Certification Organization helps those with coeliac by ensuring gluten-free standards in their certified products. In the United Kingdom, the crossed grain, a licensed symbol for gluten-free certification that is extremely useful on special products for coeliacs, has become a multibillion-dollar business, largely because some consumers see it as added value, even on products that are naturally gluten-free.

If you tune in to popular wellness trends, almost any food today has the potential to scare and is guaranteed to make you either sick, addicted, or fat, or, at worst, lead to dementia. Biting into an apple without washing it first, the way our grandparents used to do, is unthinkable for many today. Before we eat anything, we worry about its effect on our health. You may have wondered: Am I allowed to eat this egg? Yes. Why? Your body needs the cholesterol to build cells. Is fructose better than sugar? Not necessarily. Too much of it is harmful to your cells and can lead to a buildup of fat in the liver. Are yellow and red carrots genetically modified? Carrots used to come in all kinds of colours. Our modern carrot is the result of selective breeding. Many suspect that the reason it's orange is that in the seventeenth century, the Dutch (the leading carrot producers at the time) wanted to dedicate it to Prince William of Orange. Do potatoes make you fat? That's up to interpretation, but they have a regular place on the menu across all the so-called Blue Zones—geographical areas with the highest number of centenarians. Is coffee good for you or bad for you? Both, depending on whom you ask. Researchers Jonathan Schoenfeld from Harvard University and John Ioannidis from Stanford University ran the ingredients from a random selection of recipes in a cookbook through the medical search engine PubMed. They found that for nearly all of their selected ingredients—including coffee, flour, butter, eggs, milk, sugar, salt, olives, cheese, beef, and wine—there were conflicting studies suggesting they could be both good and bad. In fact, they found, 'Associations with cancer risk or benefits have been claimed for most food ingredients.' In a world of clickbait headlines, it would seem that *everything* causes cancer and that *nothing* does. Is this surprising? Anyone can go out and scientifically 'prove' their theory and add their two cents to the running of the food-rumour mill.

What about so-called superfoods? Is it surprising that açai, goji berries, and chia seeds sell so well, given that they possess 'proven' healing powers? The product labels of these superfoods read like pieces dreamed up in a writing workshop: Once upon a time in a faraway land there lived a people who were incredibly strong and who lived very long. And then one day a passing traveller accidentally discovered their secret: a seed. He ate the seed and was cured of a disease, whereupon he founded an initiative for local farmers who would receive 10 per cent of the absurdly high sales price. It is hard to imagine that superfoods are farmed on an industrial scale or that they are treated with pesticides or irradiated against bugs before being shipped, but it's true. Domestic linseeds are just as rich in omega-3s as foreign-grown ones, and picking blackberries by the wayside saves you money and helps you immerse in nature, not to mention the benefits of eating local; they are, in all probability, more nutritious than the farmed varieties from Peru.

FOOD FOR THOUGHT: It takes only an eighth of a gram of wheat to cause digestive problems—if you have coeliac disease. If you don't, then having one pizza-free day a week isn't going to help you live longer or lose weight. Don't be so quick to cut out gluten at the recommendation of your best friend's brother's girlfriend. Having a bagel or a slice of pizza may not be the worst thing for you—just don't have them every day.

CHAPTER 9

Carb Phobia

How do carbohydrates help you survive?

IMAGINE YOU ARE GOING TO SPEND a year on a desert island, and you can take only water and one other food with you. Choose the food that you think is most likely to ensure your survival:

a) corn
b) alfalfa sprouts
c) hot dogs
d) spinach
e) peaches
f) bananas
g) milk chocolate

Which one do you choose? This is exactly the question the psychologist Paul Rozin asked a group of Americans a few years ago. Bananas fared best, chosen by 42 per cent of respondents, followed by spinach (27 per cent), corn (12 per cent), alfalfa sprouts (7 per cent), peaches (5 per cent), hot dogs (4 per cent), and milk chocolate (3 per cent). This means that overall only 7 per cent of people chose a food that would actually stand them in good stead to survive on the island: hot dogs and milk chocolate. Why those two? Hot dogs and milk chocolate contain high amounts of fat and carbohydrates—a feast for the body's energy stores. Living on a desert island is all about survival, not losing weight. At most, bananas provide a short-term high thanks to the amino acid tryptophan, a building block of the happiness hormone serotonin. The fact that spinach scored so high is likely due to Popeye the Sailor Man, who developed superhuman strength after eating it by the can in the eponymous comic strip.

Most of us would probably do the same and intuitively opt for one of the healthy, nutrient-rich foods. The result of the study shows how inclined we are to label foods as 'healthy', 'unhealthy', 'harmful', 'harmless', 'allowed to eat', and 'not allowed to eat'—sometimes without good reason. Take fat for example. According to Rozin, people have come to consider even small amounts of fat as some kind of poison, even though it is vital for life. Not to mention the fact that all fats are not created equal, and that it is important to differentiate instead of hearing alarm bells ringing and thinking of weight gain and cardiovascular problems. 'Worrying about food is not good for your health,' Rozin says.

Really, it isn't. Under the constant bombardment of diet advice and food taboos, serenity and enjoyment have been replaced by fear and uncertainty in many kitchens. In his book *In Defense of Food: an eater's manifesto*, bestselling author Michael Pollan puts it in a nutshell: 'Thirty years of official nutritional advice have left us sicker, fatter, and more poorly nourished.' That is our dilemma.

FOOD FOR THOUGHT: We've been told again and again that carbohydrates (and sugar) are bad for us—but they're actually fuelling our bodies. Now, what happens when you cut out carbs altogether? Be prepared to feel lethargic, dehydrated, and possibly constipated (because foods high in carbohydrates are generally also a good source of fibre, helping along your digestive system). On a desert island, the cards would be stacked against you. You'd be staring glumly at your alfalfa sprouts rather than foraging for food or hunting animals: You wouldn't have the energy.

CHAPTER 10

'Lose Ten Pounds Fast!'

You've tried every *diet—*
why don't any of them work?

It is quite likely that you can list a number of failed diet attempts in your food autobiography. It is equally likely that the fire of your initial enthusiasm was soon extinguished, only to be replaced by a sense of frustration. The slim-in-thirty-days bikini diet didn't work for you? The low-carb and high-protein regime didn't either? Even the lemonade diet had no effect? You're such a failure!

That's nonsense of course, and fortunately there's definitive proof. In 2015 an elaborate study conducted by scientists at the Weizmann Institute of Science in Rehovot, Israel, which involved eight hundred volunteers, caused a big stir. The participants had their blood sugar levels measured every five minutes for a week. An app on their phones also logged detailed information about their eating and sleeping behaviour as well as other physical activities.

The result: 'General recommendations are not always helping people,' according to biologist Eran Segal, one of the researchers at the Weizmann Institute. What is healthy for one person doesn't necessarily work for another. Everyone is different. This sounds obvious, but nutritional science tends to treat everyone the same. 'Sometimes the participants responded in exactly opposite ways to identical meals,' said Segal. The vast differences among individuals had not been given enough consideration by nutritional scientists in the past, he added, and he found huge gaps in the research. For example, one participant saw her blood sugar level soar after eating tomatoes, while others saw their levels increase more significantly after eating sushi than after eating ice cream. Could it be that a chocolate chip muffin might be the best breakfast food after all?

How is that possible? According to scientists, it's partly due to the microbiome in our gut, as well as to our age, BMI, and of course the level of our physical activity. To control elevated blood glucose levels, they say, we don't need general dietary recommendations but rather personalised nutrition advice. 'Maybe we're really conceptually wrong in our thinking about the obesity and diabetes epidemic,' Segal said. 'The intuition of people is that we know how to treat these conditions, and it's just that people are not listening and are eating out of control—but maybe people are actually compliant, but in many cases we were giving them wrong advice.'

On the one hand, that seems an outrageous statement. On the other, it just goes to illustrate the history of science. Scientific findings are never written in stone. Foods that only yesterday were deemed taboo and damaging to our health may well be considered harmless tomorrow. Again and again we find that foods are regarded in a new light—think coffee, and whether it's a diuretic or not. Carrots, we learned, were never able to *improve* eyesight (though they are beneficial to eye health), a misconception promoted by the British army during World War II to conceal from the Germans the real source of their ability to see at night—all-new radar technology.

True, one-size-fits-all weight-loss plans seem the perfect fit for a society striving for optimisation, but regarding their actual substance they're usually not worth the paper they're written on. The psychologist Traci Mann tears such generalised plans apart in her book *Secrets from the Eating Lab: the science of weight loss, the myth of willpower, and why you should never diet again*, an attack that has made her many enemies among diet experts. Mann doesn't mince her words and opens her first chapter with these: 'Diets don't work.' Take our genes, for example:

There is a close relationship between our genetic makeup and our (excess) weight. If your parents are overweight, you're more likely to be carrying excess weight, too, than if you had thin parents. According to Mann, 'Genes play an indisputable role in regulating an individual's weight: Most of us have a genetically set weight range. When we try to live above or below that range, our body struggles mightily to adapt.'

Willpower is another obstacle on our path to a slimmer self. Temptation lurks everywhere. To be on a diet is to permanently resist that temptation and to become a master of self-chastisement. Pumpkin pie and stuffing at Thanksgiving? Verboten! Fresh pasta with creamy truffle sauce? That should be illegal! Bagel with chocolate spread? Forget about it! Belgian waffles? How dare you?! The more we deny ourselves and the more dogged we become about keeping our cravings in check, the more our desire grows. The willpower of even the most ascetically inclined people will eventually buckle under the pressure of ever-present temptation. Odysseus's strategy to have his men seal their ears with wax in order to withstand the bewitching song of the Sirens cannot be translated into our modern lives.

Ingrid Fedoroff from the University of British Columbia illustrated the traps that make self-control such hard work in an experiment some years ago. The test participants were exposed to the smell of pizza for ten minutes and then invited to eat as much pizza as they liked. Her results showed that people who are normally restrained eaters eat significantly more after such cues. As the saying goes, 'the road to hell is paved with good intentions'. Whether we are trying to save money, do more exercise, or eat more healthily, we are constantly led astray by the temptation of immediate gratification.

FOOD FOR THOUGHT: Throwing all healthy intentions overboard would be a shortsighted reaction. We are not doomed to obesity. But the pervasive talk of self-denial in the media isn't helping either. After all, the point of a diet is also its greatest weakness: Diets promise quick results, but those results rarely last (ever heard of the *yo-yo effect*?). A broader change in diet is more promising. Or, alternatively, try sticking to the Japanese philosophy *hara hachi bun me*, which roughly translates to 'eat until you are 80 per cent full'.

CHAPTER 11

Enough Is Enough

Why don't you know when you're full?

How do you know when you've had enough to eat? Is it the moment when you feel a sense of discomfort in your stomach? Or can you tell by your empty plate or the mountain of chicken bones in front of you? For most people, a noticeably full stomach is the most reliable sign that they've eaten enough. In short, they rely on a signal from a stomach that has been fairly well stretched after years of overeating. Indeed, the stomach is a hollow organ that can be trained like a muscle, which is why theoretically, at least, its volume can be doubled within a short stretch of time.

In the 1980s, Allan Geliebter conducted an experiment at Columbia University in New York in which he introduced balloons into the stomachs of his test participants. Then he pumped water into the balloons, in increments of 100 millilitres. After each incremental step, the participants were asked how full they felt. The results showed that lean people have a stomach capacity close to 1,100 millilitres, while the obese have a capacity of 2,200 millilitres or even more.

People who rely only on the fullness signals sent from their stomach wall to their brains run the risk of eating more food than is good for them. Fortunately our bodies have additional sources of information, such as the nutrient density of a food. A chocolate bar may not stretch the stomach, but it satisfies the body's need for carbohydrates and fats. An important hormone in this process is ghrelin. Released in the stomach lining, it sends signals to the brain, where it influences complex processes such as appetite, sleep, addiction, and satiety. When you are hungry, your ghrelin levels increase; when you take in food, the production of ghrelin decreases. Carbohydrates lower your ghrelin levels particularly quickly, but the levels rise again soon after. Fats, on the other hand, lower ghrelin levels more slowly and keep them low for a longer period of time, which explains why a handful of nuts keeps hunger at bay for longer than a donut.

The problem is that it can take up to twenty minutes for hormone-driven fullness signals to reach the brain and give the order to stop eating—and twenty minutes is plenty of time to eat one, two, or even three too many high-energy snacks such as a Mars bar.

Given how complicated it is to correctly interpret the signals from our own bodies, what stops us from going for third and fourth helpings? From a purely physiological point of view, nothing is keeping us from having thirds and fourths. This was shown in an experiment with amnesia patients who were offered a second meal after completing a first one. As they could not remember, they ate two complete meals in a row—apparently without feeling full. Even when they were offered the same meal a third time, they didn't hesitate to tuck in. Only the intervention of the researcher stopped them from eating a full third meal. We're incredibly dependent on the memory of past meals and are easily led by stimuli.

So does common sense keep us from constant eating after all? Yes and no. Habits play an important part as well. Generally, we eat as much as we're used to eating. Or until we've cleaned our plate. A clean plate is one of the strongest cues of all: We trust that we will automatically feel full with the last mouthful and happily ignore the fact that we might feel satisfied *before then*.

Participants in an experiment conducted by the Food and Brand Lab at Cornell University were led to eat continuously via a simple trick. Using a hose and pump connected to bowls fixed to the table, the study leader kept topping up the participants' tomato soup—and they kept slurping the never-ending supply of soup. Some continued to eat even after the experiment had finished. One participant, who had eaten three times as much as his neighbour whose bowl hadn't been kept topped up, declared the soup 'pretty filling'. Unconsciously relying on the empty-plate signal, those who had eaten from the manipulated bowls consumed on average 73 per cent more than the control group.

The fact that we usually eat from regular plates doesn't devalue the significance of the study. At home we happily shuffle between the kitchen and the couch to replenish our plates with small portions. This means that we lose track of the amount of food we're consuming. When we divert our attention from what and how much we're eating to other things, such as watching television, we fall into the same trap. Participants in an experiment who played solitaire on a computer while eating felt less full afterwards than the control group who concentrated on their meal, and the solitaire players felt hungry again sooner.

Speaking of appetite, according to a study conducted in 2013, bacteria in your gut can affect your appetite, too. In her book *Gut: the inside story of our body's most under-rated organ*, Giulia Enders writes,

'Late-night cravings for chocolate-covered toffees followed by an entire bag of party pretzels do not originate in the organ that calculates our tax returns. Not our brains but our guts are the home of gangs of bacteria that crave hamburgers after three days on a diet.' And what is more, bacteria are also at play when it comes to feeling full. According to Enders, 'Several studies have shown our satiety signal transmitters increase considerably when we eat the food our bacteria prefer. What our bacteria prefer is food that reaches the large intestine undigested, where they can then gobble it up. Surprisingly enough, those foods do not include pasta and white bread ;-).' Much better options are potatoes, endives, garlic, onions, and parsnips.

In any case, in order to regain control of your calorie intake, it's advisable to eat mindfully. Not on the go, but at the table. In her book *First Bite: how we learn to eat*, Bee Wilson pleads for more mindfulness: 'If we are going to change our diets, we first have to relearn the art of eating.' It is impossible, she explains, to develop a healthy relationship with food as long as we ignore the signals of our own bodies and instead listen to external cues such as portion size to tell us when we're full.

FOOD FOR THOUGHT: Do you tend to clean your plate every time you eat? Most of us do, and it's a habit we would benefit from breaking. Pay attention to your body. Do you feel full after only half a portion? Save the rest for later! Don't eat while you're distracted by work or television. It sounds more complicated than it is. Sometimes a little trick such as reducing the size of your plate is all it takes. And sometimes it's better for food to go to waste than to your waist!

CHAPTER 12

Love, the Anti-Diet

Is your partner making you fat?

AS THE SAYING GOES, WHEN YOU want someone to fall in love with you, the way to their heart is through their stomach—but this can also be a way to fall out of love. That may be a bitter pill to swallow, but it can't be denied. 'It always starts with the little things,' says Danny DeVito in the trailer for *The War of the Roses*, a sinister movie in which Kathleen Turner and Michael Douglas play Barbara and Oliver Rose, a couple warring with each other to the death. While there is magic at the beginning of every relationship, the ending can sometimes resemble war.

These 'little things', which suddenly begin to drive us mad, often include our formerly much-loved partner's eating habits: the way he holds his fork, the way she tilts her head, the noise she makes while chewing, the way he picks at a chicken drumstick, the way she peels an apple, the way he wolfs down his meals as if it were a race. The mannerisms we found kind of cute or even exciting when we were first getting to know each other have now become repulsive. Once the veil has been lifted, the big guns come out, and feelings are acted upon. In the film, an extreme scene takes place one evening in the Roses' home: The lady of the house is entertaining, her guests happily gathered around the table waiting for the main course, when Oliver walks into the kitchen and coolly pees on the fish his wife was about to serve for dinner.

That obviously goes too far, but it illustrates how emotional we can become where food is concerned. Think of couples newly in love. They liberally share their food across the restaurant table—of course she can try his steak and fries, and he her pasta and pie. Enraptured, they'll hold up their forks and feed each other. What's mine is yours, and so on. If our new partner likes fish, we suddenly like it, too, and if he has a fondness for pâté we might even overcome our own disgust and give kidney or liver a try. Eating together unites us—so much so that women often start caring less about their diet when they move in with their partner. It's as if they think, 'That's it!'—love has been secured, and so love handles must follow. This dynamic was explored in a study at the University of Newcastle in Australia that looked into the eating habits of couples. And how are men affected by such a change in their living arrangements? Positively so, it appears. In heterosexual relationships, men tend to benefit from the influence of their usually more health-conscious female partners. They eat more fruit and vegetables and devour less frozen pizza and beer on the couch in front of the television—contrary to the typical bachelor.

So, men in a relationship eat more healthily and gain an advantage as far as weight is concerned, and the changing dynamics can create a considerable potential for conflict. The French sociologist Jean-Claude Kaufmann suggests in his bestselling book *The Meaning of Cooking* that the dining table is where relationships are formed, and it is where the true state of our relationship is revealed. A meal is like a barometer that indicates how well things are going for a couple. If times are hard, things can come to a head at the table. He goes on to suggest that having to sit facing each other brings out the best and worst in us. Sitting together at the table is forced closeness, which is why we tend to have our arguments at mealtimes.

To ensure blissful mealtimes we know to avoid difficult subjects such as politics, religion, the in-laws, and relationship problems. Even so, sometimes this isn't enough to counteract the tension that has developed in some long-term relationships. A sure sign that a relationship is in crisis is when one partner always takes their breakfast eggs with them when going into the kitchen to fetch more coffee—for fear that the other would eat them. Once the love has gone completely, we reach for other methods of communication: According to the cultural scientist Walter Leimgruber, native women in South America would indicate their intentions of leaving their husband by no longer cooking for him. He, on the other hand, could imply that he wanted a divorce by refusing to eat the food she prepared for him. Nowadays we might try switching on the TV in order to defuse an escalating situation at dinner, but it won't work. The iciness of silence is as formidable as the heat of a verbal assault.

There is nonetheless another way to prevent tensions from arising in the first place. Jean Anthelme Brillat-Savarin, a French lawyer and noted gastronome of the eighteenth century, wrote about this in his masterpiece *The Physiology of Taste*. He spent twenty-five years writing his book before it was published, to great critical acclaim, in Paris in 1825. In it he says:

> When gourmandism is shared, it has the most marked influence on the happiness which can be found in marriage. A married couple who enjoy

> the pleasures of the table have, at least once a day, a pleasant opportunity to be together; for even those who do not sleep in the same bed (and there are many such) at least eat at the same table; they have a subject of conversation which is ever new; they can talk not only of what they are eating, but also of what they have eaten, what they will eat, and what they have noticed at other tables; they can discuss fashionable dishes, new recipes, and so on and so on.

He added that it was of course well known that intimate table talk such as this had a charm of its own.

FOOD FOR THOUGHT: Don't be too casual about shared mealtimes. We share most of our meals with loved ones; consider how that differs from sharing meals with friends, coworkers, or extended family. In relationships, no matter how intimate your table talk is, sooner or later mealtimes—including organising them, planning, shopping, cooking, etc.—end up as perilous terrain. Maybe love handles aren't too big a price to pay for blissful mealtimes after all, as long as you both have them.

CHAPTER 13

The Doggie-Bag Paradox

Why should you think twice before taking your leftovers home?

IMAGINE YOU AND YOUR FRIENDS ARE sitting around a table having dinner in a midrange restaurant. You've already enjoyed a substantial starter, and now you're struggling to finish your main course. You've managed to eat only half of the pasta mountain on your plate. Will you force yourself to overeat, or will you ask the waiter to pack up the rest so you can take it home? Do you ask for a doggie bag for your delicious spaghetti?

You consider a number of factors before making this decision: What are the others at the table doing? Are you the only one who would like to take the leftovers home? How important is it to you what others think of you? How afraid are you of being labeled as stingy and uncultivated? In other words, are you looking inward or outward? Do appearances matter more than your own preferences? Another major factor is the country you're in. In America—where portion sizes tend to be gigantic—the waiter will probably offer you a doggie bag before you ask. America is regarded as the origin of the doggie bag. Rumor has it the doggie bag was invented in California in 1943, when food was rationed during World War II. *The Oxford Companion to Food* cites a rhyme printed on a food container from that period: 'Are you happy over dinner? / Don't have all the fun alone. / Remember the pup who's waiting / And take him a luscious bone.' Officially at least the container was meant for your hungry pooch at home.

While commonplace in America, doggie bags are frowned upon in other countries, usually for historical and cultural reasons. In France, for example, according to the sociologist Jean-Pierre Corbeau, it used to be customary in middle- and upper-class circles to leave a little food on the plate to underscore the fact that food wasn't scarce. The lower classes, on the other hand, would eat up everything on their plates, and children were taught to do so from a young age. Today in France, where servings don't have frightening proportions and eating out is not as widespread as in America, doggie bags are simply unheard of and leftovers are not regarded as potentially reheatable meals. Nonetheless, the gourmet country is currently undergoing a kind of reeducation program aimed at turning doggie-bag critics into doggie-bag enthusiasts! In order to cut food waste, hotels and restaurants that serve more than 180 meals a day are legally obliged to have takeout boxes available and to actively offer these to their customers. It's not the first attempt to establish a doggie-bag culture in France, and it's probably not just the French who struggle with the idea of carrying the remnants of their *boeuf bourguignon* home in a Styrofoam box. The name doesn't help either, with its associations of chew toys and dog bowls.

In Great Britain, the doggie bag has a considerable image problem, too—discarding leftovers was once a symbol of wealth and nobility there. According to the historian Colin Spencer, leftovers were only fit to feed others—kitchen staff, for example. And during the Middle Ages, anything left over by the kitchen staff was handed out to beggars waiting in the courtyard. Spencer believes that the practice of leaving food on the plate still persists in Britain today; it signals that they can afford to be wasteful. Each year, twenty-one tons of food are thrown out in a typical UK restaurant.

Being comfortable asking for a doggie bag at the end of your meal may be as simple as thinking about it beforehand. To explain our unease about the doggie bag, psychologist Brian Wansink refers to the *endowment effect*, borrowing from behavioural economics. We tend to ascribe more value to things merely because we own them. We also have trouble letting go of them, so we try to finish everything on our plate rather than relinquishing it to a doggie bag. A simple way around this tendency is to walk into the restaurant already thinking about taking half of our food home (two meals for the price of one!). We already know that the portion size will be too large for one meal—wrap it up and take it home.

FOOD FOR THOUGHT: Get a dog, or pretend that you're looking after your neighbour's Jack Russell terrier—at least you'll have a valid reason for asking for a doggie bag. When it comes to consuming food, place your own needs over the judgement of others, and eat only until you're satisfied. Though you may have a Depression-era voice in your head saying, 'Clear your plate!' your body knows that these are times of plenty. It may be a predominately American tendency to ask for a doggie bag, but the rest of the world is catching up—if not for their own health then to reduce waste in the environment. In France, restaurateurs countered their customers' aversion to the doggie bag by simply giving it a new name: *le gourmet bag*.

PART III

How You Savour with Your Ears

CHAPTER 14

Music to Your . . . Stomach

How does the sizzle make you salivate?

WHEN YOU'RE IN A RESTAURANT AND you ask the waiter for a recommendation, you're bound to be told the dish of the day and, if it contains meat, all the details about its high quality and anatomical origin: prime rib, shoulder, tenderloin, rump, or top round. But is that enough to make your mouth water? According to Elmer Wheeler, a famous New York–based marketing guru circa 1930, restaurants would do well to drop these lengthy explanations and instead aim straight at the customer's appetite with a single noise: the sound of the meat sizzling in the pan. 'Don't sell the steak—sell the sizzle,' Wheeler recommends in his 1937 best-seller *Tested Sentences That Sell*. 'The sizzle has sold more steak than the cow ever has,' he says, although he does admit, ‘The cow is, of course, mighty important.’

Whether it’s the crackling of a fire, the chirping of a cricket, or the sound of the surf breaking on the beach—it is a scientific fact that sounds affect our appetite. The psychologist Charles Spence explains this phenomenon as a result of behavioural learning. The brain uses cues from one modality (hearing) to inform another (taste). What food do you think of when you hear a crowd cheering and the crack of a baseball bat? You’re bound to picture hot dogs and caramel corn—or are you craving a cheeseburger?

Acoustic cues don’t just affect our appetite, they also affect our sense of taste: Could you eat oysters on a chicken farm? Perhaps, but it would take something compelling to encourage you to do that, even if you *love* oysters. Spence conducted an experiment testing just that, together with the renowned chef Heston Blumenthal. They offered thirty-three volunteers two oysters, asking them to rate each in terms of pleasantness and intensity of flavour. The catch was that one was served in its shell, in a wooden basket, while the other was shelled and presented in a petri dish. While eating the first oyster, participants listened to sounds of the sea—seagulls squawking and waves crashing. While eating the second one they listened to the sound of farmyard animals. As expected, the experience of eating the first oyster excelled on both criteria as compared to the second. Appearance, sound, and the expectation of the food itself clearly did not belong together.

In another experiment, Spence and Blumenthal served their volunteers two scoops of ice cream with the bizarre flavour of scrambled eggs and bacon, but one scoop was accompanied by the sound of bacon sizzling in a frying pan and the other by chickens clucking. Depending on the background noises they heard, the volunteers rated the taste (any overall judgement aside) more bacony or more eggy—even though the two scoops were from the same batch of ice cream! But with or without background noises, one thing is certain: 'We cannot eat or drink in an environmental vacuum,' Spence said at an EMBL conference in November 2014. 'The brain just can't do it, it is constantly processing information in order to arrive at the ultimate answer: Do I or don't I like the taste?'

Another example of how pairing sounds with food can enhance our taste experience can be seen in an experiment carried out at a renowned chocolate shop in Antwerp, Belgium. Tourists visiting the shop must have thought they were in chocolate heaven when they were invited to try free chocolates created by the cutting-edge chocolatier Dominique Persoone.

Although they all ate the same type of chocolate, they experienced its flavour very differently depending on what they heard while eating it. The first of four groups of volunteers heard a Brazilian bossa nova song while eating the chocolate but were given no other context; the second didn't hear the song, only ambient noise; the third heard the song and were told that it had inspired the chocolate; and the fourth heard the song and were told the researchers were analysing the effect of its sound on the taste of the chocolate. So, how much do contextual clues here matter? As it turns out, quite a lot. The tasters who heard the song (which either triggered associations with high-quality South American cocoa beans or was simply more pleasant than listening to ambient sounds) or were told that it had inspired the 'bossa nova chocolates' rated them particularly highly: They were prepared to pay 20 per cent more for them than before the tasting.

FOOD FOR THOUGHT: Whether it's the pop of a beer can being cracked open, the spitting of steaks, or the crackling of wood, the sizzle effect explains why we love summer-night barbecues on an open fire. It's like stargazing through a telescope, says Wheeler about selling the sizzle: You may wonder where all those stars appeared from, but they've been there for centuries—you just didn't have a telescope with which to see them. He really was a marketing pro.

The power of marketing can convince anyone that Belgian chocolate (in Belgium, no less) is from Brazil. What other origin stories are you told in your day-to-day life? Are they told through intricate description? Through vivid colours? Through sounds that evoke a specific culture or vista?

CHAPTER 15

Smack-and-Slurp Phobia

How can you tell if you're a misophonic?

DOES THE NOISE OF SOMEONE SLURPING soup at the table fill you with rage? Joan Allen's character in the film *The Upside of Anger* is so angered by her guest's poor manners as he slurps at her table that she fantasises his head exploding in a gory mess. If you, too, cannot stand noisy tablemates, then you are probably suffering from an intolerance to eating sounds, a type of *misophonia*, or 'hatred of sound'. And you are not alone: According to a study conducted by Troy A. Webber and Eric A. Storch at the University of South Florida, one in five of us may experience symptoms of this noise intolerance disorder and it may affect the productivity of one in ten. The neuroscientists Pawel and Margaret Jastreboff have been exploring this phenomenon—which they named misophonia—since the 1990s and report findings of intense aversions to certain classic sounds such as chewing, the squeaking of chalk on a blackboard, or the ticking of a clock. This selective noise intolerance usually develops in childhood and is based on negative experiences associated with a specific sound. Interestingly, soft recurring sounds are more likely to cause rage than sudden loud ones. Noises like smacking or slurping, or clicking jaws, are a nightmare come true for misophonics and can cause intense physical reactions such as palpitations, anxiety, and muscular tension as well as varying degrees of aggressive fantasies.

Can you ask someone to eat more quietly just because the noise is bugging you? The experts say this isn't effective; it's up to you to cope if someone else's eating noises are your personal hell. In practical terms this means keep calm, or move to another seat. But who wants to play musical chairs on their morning commute? After all, people chew everywhere in public, even if it is only gum. For a train ride, noise-cancelling headphones are the best option, but they may not go down so well at the family dinner table.

Misophonia can be graded into clearly differentiated levels. In a self-assessment questionnaire created by Guy Fitzmaurice at Misophonia UK, the scale ranges from zero ('I know the trigger sound, but have no discomfort') to ten ('I hurt myself or others'). Level four means, 'My discomfort is such that I ask others to stop making the noise.'

According to various surveys, lip smacking is a common reason for divorce. If you don't want to lose your partner but need to put an end to the agony caused by certain noises, try one of these techniques: relaxation and cognitive behavioural therapy. These can help to reduce the sensitivity of your ears. Behavioural scientist Thomas H. Dozier suggests that we regard misophonia as a conditioned behaviour where the acoustic cue first triggers a physical reaction, which in turn leads to feelings of rage and disgust. This stimulus-response pattern, he says, can be treated with counterconditioning. The sufferer is exposed to the trigger, but only briefly and at low volume, in a comfortable environment. The duration and volume are then gradually increased, until, over time, the person learns to tolerate the sound. At least that's the hope; after all, with all the will in the world, there is no way of eating an apple quietly. And you shouldn't have to, after all; we know that the apple's loud crunch signifies freshness and intensifies the taste experience (see 'The Perfect Crisp', page 85).

FOOD FOR THOUGHT: Can't stand the sound of people eating popcorn at the movies or want to blow the head off of someone slurping soup? You're likely among the 20 per cent of people who have some degree of misophonia. It's not to be confused with *phonophobia*, which is a *fear* of sounds. A notable feature of misophonia is that it triggers a fight-or-flight response, leading many to believe it's related to psychological disorders.

If you can't stand tapping pencils or lip-smacking sounds, you may not actually *hate* those sounds; after all, they are a sign of bad manners—and we naturally distance ourselves from those who display them! If someone around you is really going at it—from clicking to sucking to slurping—it may take excessive control on your part, but you should *politely* ask them to stop.

CHAPTER 16

The Flavour of Music

What soundtrack should you play at your next dinner party?

A FRENCH RESTAURANT PLAYING SCHMALTZY ITALIAN pop tunes would be gambling with their business just as an Italian restaurant would be if it were serenading its customers with French chansons. In the same way, you wouldn't blast Metallica when you have friends over for dinner. All this is self-explanatory (think: congruence), but it doesn't fully capture the complexity of the matter. Music at mealtimes is much more than just a pleasant backdrop—it has the power to significantly alter the taste of your food.

British Airways is putting this effect to good use on their long-haul flights: They serve a playlist that suits the menu to compensate for the loss of flavour caused by the noise, dry air, and change in air pressure. The songs are played in a strict order. For starters, there's Louis Armstrong (low tones for a savoury dish) or Paolo Nutini (Scottish music for Scottish salmon), followed by Debussy or Lily Allen (both with high-toned piano to enhance sweet and bitter notes) for mains, and finally there's James Blunt or Madonna (with piano to boost sweet notes) for dessert. The airline followed the advice issued by Oxford University, which conducted a number of tests to study the effect of sounds on our taste buds. The airline's menu-design manager, Mark Tazzioli, said in an interview that in the air our ability to taste is reduced by 30 per cent, and—being the marketing pro that he is—British Airways is doing all it can to counteract the phenomenon.

Scientific trials for the pairing of music and food were first carried out in 1997. The experiment took place at The Fat Duck, the restaurant owned by Michelin-starred chef Heston Blumenthal, who is well known for his multisensory ingenuity. On the outskirts of London, patrons at The Fat Duck are served a dish called 'The Sound of the Sea'. The sea-food dish arrives together with a conch shell holding an iPod so that the customers can listen to the sound of seagulls squawking and waves crashing on the beach while they eat. It reminds them of being by the coast, and it must work since the diners all comment on the exceptional freshness of the seafood—as if it had come straight from the sea (for more on the experiment that led to the creation of this dish see 'Music to Your . . . Stomach', page 70). In another experiment at Oxford University, test subjects were given four identical pieces of 'Cinder Toffee' for dessert that contained both sweet and bitter components. They first listened to high-pitched sounds while eating the toffee, and as a result predominantly tasted its sweetness. They then ate toffee while listening to low-pitched sounds and mainly noticed its bitterness.

If high-pitched melodies can make candy taste sweeter, can they potentially elicit sweet notes from a dry wine? Apparently they can. Savoir Vivre, a wine and food trade show held in Hamburg's historic stock exchange, hosted a wine tasting accompanied by a performance of the internationally renowned chamber ensemble Trio Alba. For every piece played by the group, guests were offered a glass of wine that they were told perfectly matched the music. In reality it was always the same wine. However, its flavour changed with every variation in the mood of the music—the gentler the tones, the more harmonious the wine appeared. The change was so profound that the audience was utterly convinced that they had tasted different wines.

We know that music has incredibly powerful yet subtle effects on our psychology and perceptions, but there are whole genres yet to be explored. The music professor and composer Elmar Lampson describes hearing as a structured, meaning-generating process involving both the brain and the ear working together to actively produce the auditory sensation: 'Hearing shifts the coordinates of our consciousness; we move to another state. Not only am I hearing something—I am in an auditory space where I sense cold or warm, where there are tactile sensations and odors and also the physical impression that something is coming towards me. It's a world in which thinking and feeling become permeable with each other.' We're only beginning to understand the ways we react to auditory stimuli through our other senses. Lampson thinks that the acoustic potential is far from being exhausted: 'Music has a direct effect on the vegetative nervous system and can have a more immediate effect than images. This allows us to generate specific subconscious emotional responses.' We already know that sound directly affects our sense of taste, but what of our other senses? Can it influence what we see, smell, or feel?

To win over your taste buds it's important to find the 'right' music. Music also has a big impact on tips.

The music psychologist Adrian North observed diners' reactions to different types of background music. Whenever classical music was played, the customers spent more on starters, desserts, and coffee and were more generous with their tips. With pop music, diners would still leave tips, but noticeably smaller ones. The smallest tips were left when no music was played at all. Apparently the customers had only one thing on their minds then: Eat up quickly. It appears that when we want to treat ourselves to something special, we feel validated by classical music. Incidentally, North discovered something else: Music influences the speed at which we eat. When listening to classical music, we chew more slowly and take more time for our meal. Faster music encourages faster chewing.

FOOD FOR THOUGHT: With this knowledge, will menus in the future be set to music, as composer Richard Strauss once demanded? A steak house spicing up its dishes with, say, 'Out of Control' by the Chemical Brothers or Tchaikovsky's *1812 Overture* would indeed emphasise their hot and hearty flavours, but it would definitely not be to everyone's taste. For a safe bet, the next time you're hosting a dinner party, opt for classical music.

Still not convinced? You can try this at home and see for yourself; all you need is some dark chocolate or a cup of coffee. Listen first to Madonna's 'Ray of Light', followed by some Andrea Bocelli, and notice as sweet notes followed by bitter ones emerge on your tongue.

CHAPTER 17

The Perfect Crisp

How do you hear what you eat?

The humble potato crisp began life in an unspectacular way in a small restaurant in the holiday resort town of Saratoga Springs, New York, in the summer of 1853. Chef George Crum was working in the kitchen when his sister accidentally dropped a very thin slice of potato into a pan of boiling oil. By the time Crum recovered the stray slice, it had turned into a crunchy brown potato crisp, which the chef liked so much that he began producing them in larger quantities. The restaurant's patrons loved it, and the potato crisp was born.

As with many legendary discoveries, there's an alternate version of the story: While Crum was working as a cook at the Saratoga Springs restaurant, a wealthy travelling salesman requested crispy french fries. In an attempt to appease the gentleman, Crum made them paper-thin and served them heavily salted. The 'Saratoga chips' were utterly delicious and customers went on to rave about them up and down the country.

Regardless of the true version of events, compared to today's crisp-manufacturing processes, both versions are amazingly straightforward. Nowadays, legions of product designers devote themselves to improving and enhancing the snack. Even more than by the wide array of available flavours—salt, vinegar, sour cream, onion, barbecue, cheese, jalapeño, honey mustard, ranch, bacon (and likely anything else you can think of)—Crum would have been puzzled by the huge effort directed at perfecting the crisp's texture, available in an equally confounding number of options, ranging from straight-cut to crinkle-cut to thick-cut, and so on. 'Why?' the inventor might well have asked. Because, some might answer, the sound they make when crunched in the mouth has a lot to do with how much we like them. Determining the ideal 'crunch' is of the highest importance! We associate the sound of biting into a crisp with quality and freshness. To appeal to our notions about texture, food scientists fine-tune the ingredients, consistency, size, thickness, and cooking temperature until they have achieved the most appealing sound level and overall dynamics.

To what extent are the crunch and chewing sounds critical to our taste experience? In 2004, Oxford University professor Charles Spence ran an experiment testing just that in a soundproof chamber. He asked the participants to eat Pringles crisps, which are uniform in shape and texture, in front of a microphone, and assess each individual crisp. Their written assessments were based on their own chewing noises, which were played back to them via headphones. If the sound was loud and high-pitched, the participants attested to the greater freshness of the crisp. What none of them knew, however, was that Spence had artificially modified the sounds they heard by either intensifying or dampening them. After the participants had tried 180 crisps of the same type and same degree of freshness (after each of which they spat out the crisp and rinsed their mouth with water), their questionnaires showed that the taste experience varied significantly according to noise and pitch, on a scale from crispy to limp—from fresh to stale.

We are attracted not only to the crunch of the crisp but also to the crinkle of its packaging. The rustle of the bag affects the perceived freshness of its contents; however, it is important not to overshoot the mark in the search for the perfect sound, as Frito-Lay found out the hard way. The Texan crisp giant developed a bag, made out of biodegradable material, which at more than ninety decibels was louder than a motorbike. Not only did it earn the company massive complaints, but sales of SunChips fell by 11 per cent. They recovered only when the company dropped the noisy packaging (which they had spent years developing) and replaced it with one that rustled more quietly. In contrast, the inventor of the potato crisp didn't waste much thought on any of these subtleties. Crum failed even to file a patent for his ingenious invention.

Sometimes sounds in the environment compel us to reevaluate a food's freshness. Take the Florida-based supermarket chain Publix, for example. If you've ever been to one, you'll surely remember the sound of thunder and rain that plays overhead as water is sprayed all over the fresh produce. As the composer and entrepreneur Joel Beckerman writes, 'You might have zero idea whether this does anything positive for the food as it sits on a shelf. But it *sounds* fresh.' What else are we hearing as we go shopping and choose what to eat?

FOOD FOR THOUGHT: When you think about what you eat, you picture a food's taste, its texture (its 'mouthfeel'), its mouthwatering smell, and its exquisite appearance. You don't often hear, 'The croissant was lilting!' But you also listen to how it sounds.

Does the crunch of your cos salad determine if it's crisp or wilted? Or does a quiet table at home make for a fresher meal than a noisy restaurant dining room? Does it matter if the windows are open or shut? Some sound advice: The next time a supposedly fresh crisp tastes stale in your mouth, think about the sounds in your environment. Are you listening to loud music through headphones, which is drowning out the sounds of the crisp? Is the pitch of the crunch seemingly lowered by a high-pitched bird chirping just outside your window?

CHAPTER 18

The 'Unhealthy = Tasty Intuition'

Why does junk food sound so *good?*

Imagine that there are two different kinds of biscuits on a plate in front of you: One is a whole-grain spelt biscuit, the other is chocolate with caramel icing. Which one do you think will taste better?

You probably picked the chocolate one, and if so, then right away you've fallen into the 'unhealthy = tasty intuition' trap, as described by researchers at the University of Texas. They found that 'the less healthy [an] item is portrayed to be, the better is its inferred taste'. Likewise, when we ask, 'Why does junk food taste better than healthy food?' we've presupposed that the junk food does in fact taste better—and most people don't think twice about it.

The idea that unhealthy foods taste better, even much better, than healthy ones is not only widespread; it's also a tenet we have been taught since the cradle, along the lines of: 'Eat all of your broccoli and you can have cheesecake for dessert. And if you eat all of your peas as well, then you can have it with chocolate sauce.' Vegetables before pleasure. How are children supposed to learn to love vegetables when we put it into their minds that eating carrots and sprouts is an irksome necessity to be endured in order to get to the good bit?

Countless studies have proved that the mere *mention* that a food we're about to be served is 'healthy' lowers our taste expectations. In the aforementioned study, volunteers were asked to rate the tastiness of a mango lassi (a blended drink popular in Indian restaurants). If they were told before they tried it that it was a healthy drink, then they were significantly more likely to give it a low rating. When its high calorie content was highlighted, however, the volunteers praised the drink for its taste.

The fact that we are genetically programmed to love sugar and fat doesn't help matters—and the food industry benefits from it. They tamper with our food and cash in on our innate weakness. Steven Witherly, author of *Why Humans Like Junk Food*, talks of 'dynamic contrasts': Light and dark, sweet and salty, smooth and crunchy are particularly stimulating contrasts for our brain, he says. We just love foods with a crunchy bite that melt in the mouth. Cheese nachos—veritable cornucopias of flavour-enhancing additives such as sugar, salt, MSG, citric acid, chilli, onion, and garlic powder as well as various dairy products—are but one example of an ultimate flavour combination. Junk food tastes good and provides a quick energy boost and variety. The brain stores this information and makes all our motivational resources available in our pursuit of it. Even late at night, when the supermarkets are closed, we'll go out of our way to find a 7-Eleven to fulfill our cravings.

Under these circumstances, is there anything that can be done to disrupt our taste expectations?

The answer is yes, and an effective way is through education. Researchers at the University of Kiel in Germany were able to show that the effect of the unhealthy = tasty intuition decreases the more health-conscious people are. Nevertheless, if you believe that rationality wins over taste and think that you can promote a product simply by highlighting its health benefits, you're mistaken. According to the researchers, 'The impact of automatically activated taste associations can't even be shifted by an augmented health awareness.' The assumption to be altered, i.e., that a certain food is healthy, can't necessarily be expanded to mean that it's tasty, too. Yet, despite this new insight, there is no need to feel disillusioned. In France, surprisingly, the opposite of the unhealthy = tasty intuition applies: The French expect healthy foods to be the tastier option! Researchers at the University of Grenoble in France attribute this primarily to the fact that the French are highly conscious of quality. Instead of artificial flavourings, chefs in France tend to use more herbs and spices, fresh garlic, and shallots. They create salads from ingeniously combined ingredients, for example, lemon zest and coriander with tomatoes chopped into tiny chunks, a technique that allows the flavours to unfold the instant they hit the tongue.

FOOD FOR THOUGHT: There's no need to up and move to France just to outwit the unhealthy = tasty intuition. The first step to intuiting what's good for you is being conscious of how specific foods affect your body. Taking inspiration from French cuisine and always cooking with fresh, bright ingredients should do the trick as well.

Think about your expectations of a food's nutrient content. Is that judgement based on what it looks like and how it's presented? Its name? Packaging? Do you consider how many calories it has? Would you be surprised to learn that a chicken Caesar salad typically has over one thousand calories?

What other factors help you determine which foods are 'good' for you and which are 'bad'?

CHAPTER 19

Status Anxiety à la Carte

Why is the language of menus full of red herrings?

At the restaurant Cheval Blanc in Basel, Switzerland, awarded three Michelin stars, the menu features dishes from all over France, such as lobster from Brittany, black truffles from the Périgord, and a selection of soft and hard cheeses created by Maître Bernard Antony from Ferrette in Alsace, considered by many to be the best cheese *affineur* in France. The sardines served at the two-starred restaurant St. Hubertus in San Cassiano, South Tyrol, Italy, originate from Lago d'Iseo in northern Italy. And the menu at the two-starred restaurant Lafleur in Frankfurt, Germany, includes dishes like 'Dipped Scottish Scallops' and 'Saddle of Venison from the Odenwald'. Diners at Michelin-starred restaurants expect the best of the best food as well as a detailed, descriptive menu. Every menu is a collection of product descriptions: a sales and control tool that kindles expectations, associations, and anticipation. If you've read the many articles about the very likable 'Master of Cheese' Bernard Antony, and then see his name on the menu, you feel like you're in good company. Likewise, learning from the menu that the gourmet sardines grilled in San Cassiano were sourced locally refines their flavour by association before you've taken a first bite.

The study of the power of language is an intensively farmed field in science. Stanford University professor of linguistics Dan Jurafsky recently added a fascinating book to the mix: *The Language of Food: a linguist reads the menu*. Apart from giving us the history of ketchup, it tells of a computer program that Jurafsky and his colleagues from Carnegie Mellon University wrote to analyse 6,500 online menus from seven American cities, with prices that span the gamut from casual diner to haute cuisine. Analysing this mountain of data showed that very expensive restaurants mentioned the origin of their food more than fifteen times as often as more reasonably priced restaurants. 'This obsession with provenance is a strong indicator that you are in an expensive, fancy restaurant.'

We know that in fast-food restaurants, the most time-consuming issue is deciding what to eat, having to go through pages and pages of a (usually sticky) laminated menu. Endless lists of meals to choose from, in large or small portions, with one side dish or more: the H&M of the catering industry as it were, as opposed to the minimalistic gourmet restaurants that feel more like Prada boutiques. What we know from experience, Jurafsky illustrates with facts: 'We found that expensive ($$$$) restaurants have half as many dishes as cheap ($) restaurants, are three times less likely to talk about the diner's choice, and are seven times more likely to talk about the chef's choice.'

Lower- and midrange restaurants tend to use filler words such as *delicious*, *golden brown*, *crispy*, *rich*, *tender*, *soft*, *fresh*, and *zesty*. But why should you need to highlight, for example, that the salad is crisp? Doesn't that go without saying? Apparently not. The linguist Mark Liberman suggests that we see this overemphasis on certain properties as an expression of a kind of 'status anxiety'. Midrange restaurants, he says, are worried that their customers might have doubts about the quality of their products, which is why they reassure them in black and white that everything is fresh, crisp, and delicious—just to be on the safe side. Another favourite trick is to play the nostalgia card by using words such as *homemade*, *grandma's*, and so on. Studies have shown that words describing flavours trigger an increased activity in the primary and secondary gustatory cortex, the part of the brain that processes the information sent by the taste receptors on the tongue. Just as thinking of a great dish can make your mouth water, so can simply reading the words.

We know that the weight of terminology matters, but does the amount of words in the description matter? In his book *Predictably Irrational*, Dan Ariely, a psychologist and behavioural economist from Duke University, broaches this question in considering how our expectations affect our decisions. He offers a thought experiment, inviting you to imagine that you need a catering company for your daughter's wedding:

> Josephine's Catering boasts about its 'delicious Asian-style ginger chicken' and its 'flavorful Greek salad with kalamata olives and feta cheese.' Another caterer, Culinary Sensations, offers a 'succulent organic breast of chicken roasted to perfection and drizzled with a merlot demi-glace, resting in a bed of herbed Israeli couscous' and a 'mélange of the freshest roma cherry tomatoes and crisp field greens, paired with a warm circle of chèvre in a fruity raspberry vinaigrette.'

Although it may be unclear which caterer's food is actually tastier, Ariely claims, 'the sheer depth of the description may lead us to expect greater things from the simple tomato and goat cheese salad'.

An intriguing alternative comes at Alinea, a restaurant in Chicago awarded three Michelin stars, where they've done away with flowery language altogether and rely on punctuation to promote their lamb: 'Lamb .... ????? .... !!!!!!!!!!!!'

Let's just assume that it tastes great.

FOOD FOR THOUGHT: The language of menus is carefully crafted to appeal to our appetites. But we've also found that telling an origin story appeals to our emotions (see 'The Priming Effect', page 110). What is the origin really telling us about the food? Is it any tastier? More authentic? Is it best left to the imagination? Contrary to the trends that Jurafsky has documented, more and more restaurants are taking a minimalist approach with little to no description of the dishes, which make up a tasting menu that can cost two hundred dollars or more. This development raises the question of why celebrity chef Michael Voltaggio has coyly noted that it's better to 'under-write [the menu] and over-deliver on the flavor of the dish'.

PART IV

How You Think with Your Stomach

CHAPTER 20

'How Can You Eat *That*?'

Why do some delicacies delight and others disgust you?

How would you like to try a stuffed guinea pig? Or a barbecued rat? How about fried scorpions, or maybe *casu marzu*—a Sardinian specialty, literally 'rotten' cheese, containing live maggots that is considered to be extremely delicious and available only on the black market? Or how about the Swedish delicacy *surströmming*, fermented herring with an odour so awful and overwhelming that it can make you faint when you open the tin? A Filipino restaurant in New York's East Village named Maharlika has a dish called *balut* on the menu. These are fertilised duck eggs bred on a massive scale; regarded as a delicacy, they are believed to restore virility. The perfect egg will have been incubated for seventeen days at 108°F (42°C) in an incubator or in warm sand. The egg is then boiled for twenty to thirty minutes, which kills the embryo. But balut is also eaten raw. There's something about eating a half-gestated duck egg that seems wrong to current Western sensibilities. It breaks a Western taboo. We may think: The poor duckling wasn't allowed to see the light of day. In this part of the world, if we were to be served boiled dog we would gag, while for some diners in China it would be a delicious meal. To each their own.

What determines which delicacy will delight and which will disgust? The science and medical historian Dietrich von Engelhardt writes, 'Our food culture outlines a rough grid for each individual within which to develop their taste preferences.' As we grow up, overstepping the mark of this grid is penalised with social discrimination (someone may say to you, 'We don't eat that'). Engelhardt contends that socialisation within a food culture as well as our personal learning experiences help us internalise the grid so that even inadvertent transgressions cause discomfort and disgust.

Beyond its role in cultural differentiation and culinary likes and dislikes, disgust fulfills an important function: It warns us against potential pathogens and protects us against infection. It's crucial for survival! Omnivorous humans would be stranded without this sensorium for dangerous substances or food. When a face displays disgust—and it's the same facial expression the world over—others know to be careful. Valerie Curtis, an anthropologist and epidemiologist from the London School of Hygiene and Tropical Medicine, has been researching disgust for decades. She believes that we don't have to learn disgust; rather, it has developed in the course of evolution and is now firmly rooted in our genetic makeup. In one of her experiments, forty thousand people around the world were shown different pictures. Images of blood, faeces, carcasses, and pus triggered strong feelings of disgust in nearly everyone across all different cultures.

But back to the food: In an experiment, psychologist Paul Rozin served his students chocolate pudding in the shape of dog faeces. Many refused point-blank to eat it. The associations triggered by the similarity were enough to spoil their appetite, something that wouldn't have happened with three-year-olds, who are still resistant to disgust. Drinking apple juice from a brand-new urine-test container from a hospital or eating soup stirred with a factory-fresh comb are likely to be insurmountable hurdles of abhorrence. A spider flitting across a bowl of fresh pasta makes our stomachs turn. A dead fly floating in our cereal has the same effect. A rotten peach covered in greyish-white fuzz spoils the whole fruit bowl. Perceived impurity is contagious. Scientists who study emotion call it the *principle of contamination*. Everything that has touched, been near, or is thought to have been near the object that causes the disgust is considered contaminated. Rozin calls it the *law of similarity* and the *law of contagion*.

Most of us will have experienced firsthand how an aversion to a dish that we associate with bad memories can turn into utter disgust. The psychologist Martin Seligman coined the term *sauce béarnaise syndrome* after a personal experience: Out for dinner with his wife, Seligman ordered his favourite meal, filet mignon with sauce béarnaise. A short while later he was physically sick. His nausea and the vomiting had nothing to do with the meat or the sauce but were caused by a stomach bug. The stomach flu passed, but the aversion remained—only to the sauce béarnaise, not to the steak. For years, Seligman shuddered at the mere thought of the dish. He had been adversely conditioned.

And all of those foods we can't bear to think about? Experts predict that in the future we will be eating a lot of the foods that make us shudder today. This will almost certainly include things like artificial meat and protein-rich insects such as termites, grasshoppers, and caterpillars (already a growing trend!). We might as well get used to it now.

FOOD FOR THOUGHT: There are many foods we cannot even imagine putting into our mouths, but is our disgust rooted in cultural taboos or psychological aversions? A little of both? How would you feel if at the end of a delicious meal you found out that you had just eaten dog meat? Our personal cultural grids are easily redrawn as norms shift. Foods that once elicited aversions that seemed natural may soon become normal. There's little to differentiate a reasonable aversion from an irrational one. However, it's important to remember that disgust is a mechanism to protect our bodies. Listen to it. We may not be susceptible to ingesting pathogens, but who wants to be the one to mistake faeces for chocolate—and accidentally eat it?

CHAPTER 21

The Marketing-Placebo Effect

Why is there no veritas *in* vino?

LET'S IMAGINE YOU'RE INVITED TO A party and you don't want to show up empty-handed. You think to yourself, 'Wine always works as a hostess gift,' and so you dash off to the nearest wine shop. Standing in front of the shelves, you deliberate the options: red or white, Old World or New World, the one with the yellow label or the one that's blue? Does the cheap wine look expensive enough, or the expensive one too cheap? You've not got much time so you quickly grab a bottle from the midrange-price section. Quality comes at a price, but there's no need to overdo it either. As you ceremoniously present the bottle, you are beginning to have your doubts. Does the bottle show how much you spent on it? Your host casts a quick glance at the label, determines its price (and, with kitchen-sink psychology, how much they are worth to you), and mentally issues a verdict. And without either of you having tasted as much as a single drop of the wine, you both trust that you have judged its quality accurately.

So much for *in vino veritas*—you may have both been deceived by the label, but you don't actually know. Rest assured that even renowned experts get it wrong sometimes. When several bottles of prestigious Clos Saint-Denis, Domaine Ponsot, of 1945 to 1971 vintage were offered for auction in New York in 2008, nobody noticed that the labels were fake except for the domaine's owner, who had made a special trip to attend the auction after hearing these wines would be for sale. The problem was that the domaine didn't start the appellation 'Clos Saint-Denis' until 1982. Whether the actual taste of the wine would have been enough to convict the counterfeiters is anyone's guess.

But back to your regular wine shop where you're met with a labyrinth of choice: origin, grape, vintage, finishing process—how is your average consumer meant to know their way around? The (controversial) Parker point system promises to liberate the consumer from this dilemma. The principle is simple. Counting starts at fifty, wines with eighty-six points or more are deemed interesting, and one hundred points are awarded only to those wines that promise to become legendary, just like the inventor of the eponymous system, Robert Parker himself. A wine awarded ninety-nine points by the Parker Imperium can expect to multiply its value overnight. On the other hand, wines that don't achieve the eighty points necessary to cross the threshold into the 'above average' category won't even be worth a mention in renowned trade magazines. But what exactly is the difference in taste between a seventy-nine-point wine and an eighty-point wine?

You would think it would be simple enough for an amateur to tell house red from a fine bordeaux. The participants in a study conducted by researchers at the University of Bonn in Germany and the French business school INSEAD seemed to agree. The researchers gave volunteers three wine samples to taste, which they said cost three, six, and twelve euros each. Despite the fact that all three samples were exactly the same, the majority of the volunteers rated the alleged twelve-euro wine higher than the others. The price had massively influenced their taste experience. This is called the *marketing-placebo effect*. If it's more expensive, it must be better—a fallacy encountered across numerous product ranges and one that has some people forking over two hundred dollars for an eye cream.

What came as a surprise was that not all the volunteers fell for the marketing-placebo effect to the same extent. Of all people, it was the rational thinkers who were most susceptible to the delusion, something that could be observed in the participants' brains. Imaging procedures showed an increased activity in the prefrontal cortex, precisely that area of the brain that is responsible for making rational decisions. Less vulnerable people, however, showed an increased activity in the insular cortex, a region that mainly processes body signals. 'These brain structures and behaviour patterns are not innate. They are formed by connections that develop during our lifetimes. The way we learn is key to this,' the study leader said. People whose learning behaviour was more oriented towards external rewards were more liable to fall for product promises. Those who trusted their own instincts were more likely to be on the safe side. The senses can be trained of course, especially with regard to wine. The neuroscientist Gregg Solomon has discovered that sweetness, balance, aroma, tannin content, minerality, and viscosity become more easily identifiable with increased wine-tasting experience. The acquisition of this sort of expertise requires the less rational and more experience-based skill of being able to put flavours into words.

Incidentally, Robert Parker is famous for his precise and experience-rich wine vocabulary. He once rebuked a wine taster for using the phrase 'smells like the sex glands of a lemming'—because, he said, 'Who in the world can relate to that?'

FOOD FOR THOUGHT: Can any of us really tell the difference in taste between a cheap wine and an expensive one? Stick to buying and gifting wines that you know and like. If the recipient likes you, they should appreciate your choice. If you have yet to acquire the favour of the host, take advantage of the marketing-placebo effect and rely on Parker points, a tasteful label, or—if all else fails—the cost of the bottle to impress.

CHAPTER 22

The Priming Effect

Why doesn't Häagen-Dazs ice cream come from Denmark?

'EAT MORE FISH AND CHOCOLATE CAKE!' This was presumably—and in so many words—what the restaurant owner whose business was slow wanted to call out to his patrons. But as that would have caused nothing more than brief entertainment, the restaurateur used a rhetorical trick instead: He altered the names of the dishes on his menu. The previously plain 'fish fillet' became 'succulent Italian seafood fillet' and the chocolate cake was renamed 'Black Forest Belgian double-chocolate cake'. It raises the question: What was the famously German Black Forest doing in Belgium?

Using language this way is a promotional tool that's aimed to go into effect deep in the buyer's subconscious mind. A cue like *Belgian chocolate* is enough to trigger key associations and influence your purchasing decisions. Cognitive psychology calls it *priming*. Priming can be compared to a guessing game where the contestants are given stimuli such as a colour, tune, scent, or word that they have to put together—using their memory—to complete a picture. Whereas the simple term *chocolate cake* merely conveys basic information, the description *Belgian chocolate* carries with it an extra layer of emotional, cognitive, and sensory information. The restaurant saw food sales go up by 28 per cent after implementing these neuromarketing measures. Apparently, just alluding to the Black Forest gâteau in combination with the world's best chocolate was enough to make this dessert appear glamorous and sumptuous.

What do you think of when you hear the name Häagen-Dazs? Denmark? That's likely—except Häagen-Dazs is not a pricey Danish ice cream, but an American one—from a company founded in New York in 1961 by a Polish American couple, Rose and Reuben Mattus. So why do so many people automatically assume that it is a premium product from Scandinavia? Well, it's another classic case of priming. The Nordic-sounding product name evokes associations typical of Scandinavia in our subconscious minds: nature, freshness, design, and quality. Subconsciously we conclude that the ice cream somehow embodies these things.

What we think of as an impulse purchase is in reality the result of a highly complex process, triggered by neurological manipulation efforts that we encounter in any purchasing environment. Sometimes, a fraction of a second is enough for a stimulus to trigger a response. In one experiment, some participants who were solving problems on a computer were shown the words *Lipton Ice* for twenty-three milliseconds while working on the problems. Those who were primed were more likely than the unmanipulated reference group to choose iced tea instead of water after the test. The effect was particularly pronounced when the participants had eaten a salty snack before the test and were very thirsty. In that case, 85 per cent opted for iced tea instead of water.

'The more we find out about the brain, the more I think there is a need for a UN charter against neurological manipulation,' the psychologist John Bargh warned many years ago, although his main concern was probably not that people might be tricked into buying iced tea. In 1996, Bargh used a simple experiment to establish the enormous influence priming has on human behaviour. Two groups of participants, thinking they were taking a language test, worked on lists of words. One group was given a list relating to the subject of old age, which included words such as *forgetful*, *slow*, *walking stick*, and *shuffle*. The reference group was given a list focusing on youth, containing words such as *athletic*, *flexible*, *dance*, *party*, and *spontaneous*.

The real experiment started after Bargh had thanked and released his volunteers. He secretly timed how long it took them to walk the nine metres to the exit. The result: Those who had worked on the 'old age' list walked more slowly to the door; those who had been primed on the subject of youth were one second faster.

The *ideomotor response*—in which an idea affects action—also works in reverse: If you act like an old person, you will think in age-related terms. This is what researchers at the University of Cologne discovered when they turned Bargh's experiment around. Students who had been asked to walk around a room very slowly were later drawn to words related to old age.

In priming, the Nobel Prize winner Daniel Kahneman explains, 'Your thoughts and behaviour may be influenced by stimuli to which you pay no attention at all, and even by stimuli of which you are completely unaware.' Kahneman distinguishes between two cognitive processing systems. On the one hand there is the intuitive system, with its unbridled impulses, associations, and instinct-driven reactions. On the other, Kahneman presents the logically thinking, reasoning, conscious system, which makes deliberate choices and actively controls our thinking and behaviour.

Even though marketing efforts are aimed at the first system, we are not entirely at the mercy of our subconscious minds. We are in fact able to 'reprime' ourselves—for example, where our eating habits are concerned. The researchers Katie Mosack and Amanda Brouwer looked into how this works in detail. For their experiment, they split 124 women into three groups. The first group received information on healthy eating and was asked to keep a food diary. The participants in the second group were also asked to keep a diary, but they were encouraged to set themselves some healthy eating goals articulated in the form of identities such as 'I am a fruit eater,' 'I am a vegetable eater,' or 'I am a sugar avoider.' The third group was a control group and was asked only to keep a diary.

As it turned out, those who had identified themselves as healthy eaters were noticeably more successful at fulfilling their goals than the women in the other two groups. Those who had primed themselves for healthier eating by thinking of themselves as 'doers' ate markedly more healthily over the course of the study. 'The more one identifies with a particular role, the more likely one is to participate in role-related behaviors,' the researchers concluded.

FOOD FOR THOUGHT: The next time you can't resist temptation, don't feel bad about it. There are nearly imperceptible influences on many of our food choices. Start to pay attention particularly when looking at food advertisements, both on TV and in print. What associations do you have with the setting? The colours? The name of the product or company? And be wary when someone tries to sell you Belgian chocolate from the Black Forest.

CHAPTER 23

The Health-Halo Effect

What do you think of as 'healthy'—
and what actually is?

IN BAŞAKŞEHIR, A NEWLY DEVELOPED DISTRICT of Istanbul, the municipal government released forty-five thousand frogs to make the local residents feel as if they lived close to nature. In the food industry, frogs are used as a 'nature seal'. It's a marketing method called *greenwashing*.

Stamps reading 'regional', 'fair trade', 'all natural', 'family farmed', and 'sustainably farmed' may sound as if they denote 'organic' (a regulated designation), but almost none of these labels are protected by federal standards let alone provide an environmental guarantee of any kind. In fact, the stamp can be completely made up. The halo effect strikes again, intuitively inflating the quality of a product and leading consumers to draw sweeping conclusions.

Suddenly, 'regional' comes to mean healthy, even though all it means is the product is 'from the region', without allowing for the possibility that the orchard may border a freeway or a nuclear power plant. Researchers from the University of Bonn in Germany discovered increased activity in the reward-system part of the brain—the ventral striatum—in people choosing products with an eco-seal. The stimulation of the health-halo effect is particularly strong in people who buy organic products on a regular basis. They have become so habituated to the effect that they feel rewarded for their conscientious consumption and are thus prepared to pay significantly more: on average 45 per cent more.

Just how easily we are misled despite our better knowledge was shown by a study conducted by researchers at the University of Houston. Participants were shown two labels for products such as fruit snacks, breakfast cereals, peanut butter, and Cherry 7UP: one including keywords such as *whole grain*, *antioxidant*, *organic*, or *all natural*, and one with those words edited out. Participants were then asked which one was 'healthier'. They overwhelmingly responded that the products with labels containing the keywords were better for you. When we see these buzzwords, we believe that these products are good for our health. Even the Cherry 7UP, a product made up of water, high-fructose corn syrup, and food colouring, was rated 'healthier' when it was labeled 'contains antioxidants'. In *Food Rules: an eater's manual*, Michael Pollan aptly sums it up: 'For a product to carry a health claim on its package, it must first have a package, so right off the bat it's more likely to be a processed rather than a whole food.' That is, ketchup from a plastic bottle is 'fake', with or without pictures of tomatoes.

Even more surprising is that labels and keywords can have a placebo effect: In one experiment, researchers offered fit-forward participants the same trail mix, labeled either as 'fitness' trail mix with sneakers on the label or as regular trail mix with no fitness imagery. They then had the option to exercise on a stationary bike. The result was interesting: Not only did the participants who had been offered the nuts as a 'fitness snack' eat more of them (to the tune of fifty to one hundred additional calories), but they also put noticeably less effort into their exercise. The illusion of smarter food choices—which were actually poor choices—had an inverse effect on fitness. The study found that for those who want to manage their weight, 'fitness branding in food marketing can exacerbate this problem [of unsuccessful weight control] because fitness cues make eating dietary permitted food compatible with weight control'. We're tricked into thinking that eating the 'healthy' food is helping us achieve our goal. Furthermore, it's surprising that 'increased consumption of fitness-branded food may even serve as a substitute for actual physical activity'. We think we're making healthy food choices, and so don't do the exercise we need to actually be healthy.

Participants who had mentioned in the study questionnaire that they had weight issues and wanted to lose weight were particularly susceptible to the fitness effect. 'People already know to avoid cliché products such as chips and fries, but the addition of the word "fitness" adds a kind of halo to products that are not perceived as taboo per se,' the leader warned. This was important, he added, as it was counterproductive to the objectives these people had set themselves.

FOOD FOR THOUGHT: We know that words are a powerful source of influence—just as much when it comes to labelling our food. Familiarise yourself with nutrition-fact labels and find out just how many calories you're consuming. Many of those 'low-fat' or 'sugar-free' snacks are loaded up with other nasties. A quality seal alone does not make a food natural or organic, just as a few frogs do not make a pond—let alone a nature reserve. Once we become accustomed to the frogs and the quality seals, just seeing them can trigger a sense of wellbeing.

CHAPTER 24

The Romeo-and-Juliet Effect

Why does absence make your cravings stronger?

DO YOU SOMETIMES FEEL THAT PEOPLE close to you are doing exactly the opposite of what you have asked them to do? That they deliberately leave the milk out, don't clear up after themselves, play the music a tad too loud, or don't eat their vegetables? There's no point wasting your energy in getting angry; this rebellious behaviour has much less to do with you personally than you may think. It is in fact caused by a psychological phenomenon called *reactance*. We like to decide for ourselves what we should or shouldn't be doing at any given moment; requests, limitations, and bans inevitably lead to mental resistance. In short: We object. Since Romeo was banned from loving Juliet, he found her all the more desirable, which is why reactance is sometimes referred to as the *Romeo-and-Juliet effect.*

Reactance is a problem not only for husbands and wives, parents and children, teachers and students, and employers and their staff; it can also lead to the failure of political interventions. One such failure occurred in New York City in early 2013. Mayor Michael Bloomberg was concerned about the dramatic increase in obesity and diabetes among residents of his city, a trend he blamed in part on the soaring consumption of sugary drinks. So he started to look for ways of making New Yorkers drink less soft drinks. Bloomberg's idea seemed simple yet clever: If there was a ban on the sale of supersized drink containers, then people would automatically drink smaller quantities and consume fewer sugary drinks overall. In the end the ban on XXL-sized cups and bottles was never put in place. The idea was met with massive protests from industrial associations and lobbying groups, while behavioural economists including David Just from Cornell University feared that people would rebel against such arbitrarily imposed restrictions. Just argued that instead of promoting healthy eating habits, the measure would backfire as consumers and retailers would find ways to circumvent the ban—for example, through bundling medium-sized cups in two-for-one deals—ultimately buying and selling even more soft drinks. Nobody likes being patronised.

Equally, the organisers of a campaign to introduce plain milk in schools were surprised by the unexpected response to their well-intentioned plan. When a ban was introduced on all sweetened or flavoured milk such as chocolate or strawberry, the children, who used to love these drinks, started throwing out 30 per cent of the plain milk purchased, and many stopped drinking milk altogether. The ice is dangerously thin where nudging is concerned, because such behaviour control usually works only as long as it doesn't limit the range of choices or impose outright bans, but merely nudges the consumer towards making the 'right' choice (see 'Nudging', page 143).

You may also notice that self-imposed restrictions lead to tunnel vision. Are you on a low-carb diet? Are you upset that the aroma of freshly baked baguette is even more tantalising than usual? That's a perfectly natural reaction! Everywhere you look (and smell), all you notice is the banned substance that now appears disproportionately more attractive than ever before.

Just how subtle the reactance effect can be was shown with the following experiment: The participants were shown two protein bars with the same number of calories, presented as a candy bar and a health bar. Half of the participants were allowed to choose freely between the two; the others were told to have the 'healthy' one. Those who were asked to eat the apparently healthy bar felt hungry sooner afterwards. Those who chose freely felt fuller for longer. Given these findings, are health-promoting measures at all enforceable? They are, as long as they offer the right incentives.

The entrepreneur and game developer Jesse Schell, who researches psychological incentives for computer games, distinguishes between 'wannas' and 'haftas'. The difference between things we want to do and things we have to do is 'the difference between work and play, slavery and freedom, efficiency and pleasure', Shell explained at the 2013 DICE Summit, a gathering of leading video game executives in Las Vegas. Unfortunately, many well-intentioned interventions are doomed because they feel like 'haftas'.

According to Schell, the trick is to turn a 'hafta' into a 'wanna'. Social incentives can be helpful in achieving this. The desire for approval is one of the strongest motivators for action. One guarantee for a successful intervention is for peer groups to encourage the desired behaviours in each other; another is the preservation of free will. A French study involving eighty volunteers established that people are twice as likely to do something when they are reminded that it is their free choice whether to accept or refuse. As Jack W. Brehm, founder of the reactance theory, pointed out in 1966, emphasising the voluntary aspect has great impact on the targeted behaviour.

And if none of that makes any difference to your behaviour, it may help to stir up some new desires. When the French pharmacist Antoine-Augustin Parmentier planted potatoes, a tuber much eschewed by the peasant population at the time, and had them guarded on the king's fields by soldiers during the day, the thieves crept up in the night to steal the supposedly valuable crop. News of the 'precious' potato spread like wildfire, and people started growing their own potato crops in secret. Catherine the Great of Russia was said to have had her fields protected by a fence and to have made the theft of her potatoes a punishable offence. Of course, her people contrived means and ways to get ahold of the newly desirable tuber one way or another.

FOOD FOR THOUGHT: We may crave what we can't have, but are there good reasons to hold back? Consider the daily injunctions we impose on ourselves. Are there reasons not to have ice cream for dessert after every dinner? Or to restrict eating out?

Banning something altogether only makes it more attractive, as Romeo and Juliet did so much to prove. Do you have the self-control to consume treats in moderation?

CHAPTER 25

Sleep Yourself Slim

How does what you eat affect how you sleep?

SLEEP PULLS ALL THE STRINGS: The quality and quantity of our sleep determines what the next day will bring, if we will face the morning with bright or with tired eyes. We spend a good third of our life in bed—holiday lie-ins included. We ought to get between five and nine hours of sleep a day, including naps. Mental exercises such as counting sheep to help us fall asleep are counterproductive as they strain the brain instead of helping it to relax. Tradition has it that warm socks and hot milk with honey will help, and some swear by that. But it's not enough to automatically guarantee a good night's sleep. After a drunken night out, a glass of warm milk with honey won't counteract the effects of alcohol. True, alcohol helps us fall asleep faster, but it also prevents you from sleeping through the night. We may not live to sleep, but as we make our bed, fall asleep, and sleep through the night, so we must live.

The gastronome Brillat-Savarin summarised it neatly: 'Whether man sleeps, eats, or dreams, he is yet subject to the laws of nutrition and to gastronomy.' The quantity and quality of our diet has a decisive impact on our work, repose, sleep, and dreams, he claimed. Foods that Brillat-Savarin recommends eating at bedtime include dairy dishes, poultry, the flowering common purslane, and specifically reinette apples such as Cox's orange pippin (common in the United Kingdom). Certain foods and activities promote dreams: 'In general, all stimulent [sic] food excites dreams, such as flock game, ducks, venison and hare. This quality is recognised in asparagus, celery, truffles, perfume, confectioneries and vanilla.'

Eating a ham hock for your evening meal is a sure guarantee of a night spent tossing and turning. Not only does the high fat content of your dinner make it harder to fall asleep in the first place, but it also prevents you from getting a restful night's sleep. In a study published in the *Journal of Clinical Sleep Medicine*, researchers were surprised to learn just how little it takes to disrupt our sleep. They conducted an experiment with healthy participants between thirty and forty-five years of age who slept between seven and nine hours a night. During the first four of a total of five days in the sleep laboratory, each participant was given a controlled diet of fibre-rich foods with little saturated fat. On average, it took both the men and women seventeen minutes to drift off to sleep. On the last day of the experiment, the volunteers were allowed to eat whatever they desired. They didn't go for the healthy options. The food they ate contained less fibre and more fats and sugars. Suddenly it took them twenty-nine minutes to fall asleep on average. The study concluded that a single day with more fat and less fibre had an impact on sleep patterns. We could maybe get by if those extra twelve minutes were the only downside of gluttony, but they aren't. The onset of slow-wave sleep, or deep sleep, the type of sleep that is so important for our brains' recovery, is also delayed by a high-fat diet.

But it isn't only that what we eat affects how we sleep; how we sleep also affects what we eat. In a nutshell, one might say: The shorter the night, the bigger the waistline. Lack of sleep increases appetite and reduces self-control. Sleep researcher Jürgen Zulley explains, 'In our sleep, the hormone leptin is released and signals the body that it is full. That's how the body manages to cope for ten hours without food, something it wouldn't be able to do during the day. If we don't sleep, or don't sleep for long or very well, leptin's counterpart, the hunger hormone ghrelin, gains the upper hand.'

Chronic sleep deprivation can lead to obesity. Researchers at the New York Obesity Nutrition Research Center carried out a study comparing the eating habits of well-rested and sleep-deprived participants. While lying in a CT scanner, members of both groups were shown images of various food items. The brain scans revealed that the reactions to the images were much stronger in the reward systems of the sleep-deprived participants (who had slept for only four hours the night before) than in those of the normal (nine-hour) sleepers. Moreover, researchers have found that sleep-deprived people consume three hundred more calories per day than those who are well rested. The kilos pile on quickly.

People who suffer from night-eating syndrome (which has been found to be a partly genetic eating disorder) are compelled, as the name suggests, to eat during the night; tormented by hunger attacks, they raid the fridge while others are fast asleep. Those who lack sleep also lack leptin, the hormone that makes us feel full. Studies at the University of Pennsylvania have confirmed a link between hunger and sleep duration. The scientists who studied the eating patterns of normal, long, and short sleepers found that short sleepers had the highest calorie intake and consumed fewer vitamins than normal or long sleepers.

FOOD FOR THOUGHT: If you're finding it difficult to hit the hay at a decent hour, try evaluating your diet. You may feel a boost of energy from a bowl of ice cream before bed, but it may be preventing you from getting that much-needed good night's sleep. And the cycle repeats: When we haven't slept well, we reach for high-fat and sugary foods, which cause us not to sleep. Try giving yourself an extra half hour in bed every now and then—you'll make much better food choices after a good night's sleep!

CHAPTER 26

The Feeding Clock

How can eating help you beat jet lag?

IMAGINE THAT YOU'RE ABOUT TO FLY to overseas for a conference and you've got a ten-hour flight ahead of you. In economy. You shudder at the thought of it, not just because you're afraid of flying, but also because you simply cannot bear the cramped conditions. On top of this, you know that your physical distress won't improve once you've reached your destination: For the next few days, you are likely to suffer from headaches, fatigue, insomnia, dizziness, and, above all, digestive problems. The more time zones you cross, the worse your jet lag gets. When flying west with the sun, stretching out your day, you won't be hit as hard by this topsy-turvy feeling as if you were flying east, but you will still feel pretty rough. Regrettably, no jet-lagged traveller manages to look even half as beautiful as Scarlett Johansson did in *Lost in Translation*, playing an insomniac sitting by the window in her Tokyo hotel room, watching the night fade away. Bill Murray's face, crumpled from sleepless nights spent tossing and turning in bed, is significantly closer to the truth. So what can you do?

Many strategies have been designed to rebalance the body on landing. For years, people focused on light as the most important factor in regulating our circadian rhythm (our internal clocks), which is why many travellers—afraid of light shock—don their shades at baggage claim, even if it's nearly sundown. But there's a second important factor, the *feeding clock*. After studying mediating factors to the circadian rhythm in mice, scientists discovered that we're regulated not only by the light we see, but also by when we eat. When the researchers changed the feeding times for the mice, the bacteria in their gut became unbalanced. If the same goes for us, each shift in mealtimes throws our internal clock out of whack. The good news is that we can reset our internal clock by adjusting our eating time.

In the early eighties, biologist Charles F. Ehret of the Argonne National Laboratory in Illinois developed an anti-jet-lag diet, which needs to be started several days before the journey (east or west). The crux of the diet is alternating feasting (meat) days and fasting (salad) days. Caffeine is allowed only between 3:00 PM and 5:00 PM. The underlying idea is that the body is already confused before the journey has even begun. It's an intensely stressful strategy for the body; nevertheless, Ehret's 1983 book *Overcoming Jet Lag* sold like hotcakes. His son once mentioned in an interview that his father had been inundated with calls from people planning trips for everyone from President Reagan to Aerosmith.

Later, in 2004, scientists at Harvard modernised Ehret's diet and simplified it radically: No food at all is allowed from twelve to sixteen hours before landing in the new time zone. However, it is recommended to drink large amounts of fluid during the flight, but only water. The body is tricked into abstinence mode and readjusted at the destination—just like your watch. During the first few days it is also recommended to eat food rich in carbohydrates (such as potatoes, rice, pasta, and whole-grain bread) in the evenings to make you sleepy, and to drink hot chocolate instead of beer. In the mornings, carnivores should tuck into a protein-rich pork loin or a turkey sandwich; otherwise milk, eggs, cheese, or nuts will do the trick.

It is odd, however, that scientists consider fasting for hours on a plane to be harmless. As our blood sugar level drops, so does our mood, which will already be at a low point unless, perhaps, we are travelling in business class. Either way, hunger is brutal. A stomach growling for hours can make even the most easygoing person aggressive. And when the sweaty person in the neighbouring seat starts claiming our armrest, that aggressiveness can soar sky-high. After all, who would want to spend long hours with hundreds of hungry and hypoglycemic people on board a plane miles above the endless spread of the Atlantic Ocean?

Help for our appetites (and seatmates) may come in the form of the popular aeroplane drink tomato juice. Are you one of those people who have a glass of tomato juice every time you're on a plane, and only then? Ever wonder why this might be? According to researcher Charles Spence, at an altitude of 30,000, the way we perceive flavours is affected by the extremely dry air (which dehydrates the nasal passages) and the low air pressure, which is roughly equivalent to the air pressure at 2,000 to 2,500 feet on the ground. In a series of tests commissioned by the German airline Lufthansa, researchers established that 'tomato juice was rated much tastier at low pressures than at standard atmospheric pressure where it was described as stale. In the air, pleasantly fruity aromas and sweet, refreshing flavours came to the fore.' Our ability to taste and smell is dulled in lower air pressure (as if we had a cold) and so foods that are normally perceived as having a strong or bitter taste are much more pleasant in the air. This response particularly affects our perception of salt, sugar, and herbs, while umami flavours like tomato juice, which is high in natural glutamate, benefit from the higher air pressure because the great intensity of the taste we experience at sea level is weakened in the air.

FOOD FOR THOUGHT: It may seem obvious that the regularity of our circadian rhythm depends not only on when we sleep but also on when we eat. Our bodies love routine. Travelling to another time zone unsurprisingly wreaks havoc on our internal body clock, causing all-too-familiar jet lag symptoms. But it doesn't have to! Try adjusting your eating times to your destined time zone several days before you leave. Hydrating *well* may also help.

So, will regulating meal times then affect how well you sleep while travelling? Absolutely! And what about the rest of the time? Too often in our busy lives we skip meals or delay them several hours. Ever notice a direct correlation between these missed meals and a bad mood or poor sleep? If there's one thing we're particularly bad at, it's sticking to a schedule—sit down to eat and go to sleep *on time*.

PART V

How You Feast with Your Feelings

CHAPTER 27

'I'll Have What You're Having'

Why do you order what you order at restaurants?

SOME CUES HAVE AN OBVIOUS EFFECT on us while others work so subtly on our subconscious that we would never think of them as cues in the first place. Either way, we are so susceptible to manipulation that we can be sure of only one thing: Our behaviour is irrational. Nowhere more so than in a restaurant.

It should come as no surprise that the waist size of the waitstaff affects the choices you make when dining out. In a study on the relationship between a waiter's weight and customers' orders, researchers Tim Döring and Brian Wansink scrutinised the interactions between waitstaff and customers in sixty different restaurants. They recorded the estimated BMI of the waiters and the diners as well as what food and drink they ordered. The results showed that the higher the BMI of the waiter or waitress, the more food and drink the customers ordered, regardless of their own weight. Overweight waitstaff also significantly increased the diners' proclivity to order alcoholic drinks and desserts. Döring and Wansink explained that this was because the server set a 'social norm' in which they became the benchmark—against this metric, thinner diners perceived themselves as having bested the norm while heavier diners perceived themselves as the average, meeting the norm. If the server is fat, we feel justified to stuff ourselves. If the server is slim, we are more careful about how much we ourselves consume. In other words, thin waiters and waitresses personify our guilty conscience.

Even more important than the size of the waiter, however, are the eating habits of your fellow diners. If you are with a group of people who are wolfing down their meals, you automatically adapt your own eating speed to theirs. If all the others are ordering a Coke, you tend to give your usual beer a pass. If your fellow diners are overweight, you eat more. If the waiter takes everyone's order individually, the likelihood that everyone opts for a dish from the same category increases. In a study published in the *Journal of Food Quality and Preference*, Brenna Ellison found that 'diners wanted to be different from their dining companions, but not too different'. We want to fit in and not be the odd one out, which is why in company we eat more than when we're eating alone. Psychologists call this subconscious tendency to imitate the behaviour of others the *chameleon effect*. Woody Allen satirised this effect in his film *Zelig*. The opportunistic protagonist, Leonard Zelig, is a human chameleon; he takes on the traits of those around him. When Zelig is with a well-educated psychiatrist, he, too, becomes a psychiatrist; in the presence of large people, his stomach bulges.

Have you ever gone to a restaurant with a large group and been one of the last to order, only to realise shortly after that the dish you selected wasn't really what you wanted? Sometimes we're compelled to select an option in response to those around us. To demonstrate this phenomenon, Dan Ariely and a colleague approached one hundred groups of unsuspecting brewery customers, disguised as waiters, and offered them free samples of beer. They could choose between Copperline Amber Ale, Franklin Street Lager, India pale ale, and summer wheat ale. Along with the beer, they gave the customers a questionnaire asking them to indicate whether they liked the beer and whether they regretted their choice. Not all customers ordered their beer out loud; half of the groups they approached had to make their choice in writing. As it turned out, those who placed their order in public ordered a range of different varieties, an expression of individuality. Yet, in the end, they tended to be less satisfied with their choice than those who chose in private. With one exception: 'The first person to order beer in the group that made its decisions out loud was de facto in the same condition as the people who expressed their opinion privately, since he or she was unencumbered, in choosing, by other people's choices.' With the result that in the end they were happier with their choice than the other people in their group who had ordered out loud.

In other cultures where individualism is not desirable, as in Hong Kong, Ariely found the opposite behaviour to be true: 'In Hong Kong, individuals also selected food that they did not like as much when they selected it in public rather than in private, but these participants were more likely to select the same item as the people ordering before them.' According to Ariely, people—especially those with a high need for uniqueness—would sacrifice personal benefit in order to gain prestige among their peers. People make compromises in order to look good.

FOOD FOR THOUGHT: Do you find yourself ordering dessert in some restaurants more than in others? It may be that a particular restaurant has exceptionally tasty toffee crunch caramel cheesecake, but the next time you have a hankering to order it you may first want to take a look around. Do you feel the guilt of overindulging weighing on you, or might there be other weighty influences? Consider, too, the dining habits of your companions. It isn't only the waiter who nudges you to order. Ariely suggests that it helps to decide what dish you want before the waiter comes to the table. Don't let your friends' choices influence your own order. Choose a dish quickly, and stick to it.

CHAPTER 28

Nudging

How do you choose what to eat in a cafeteria?

DOES YOUR COMPANY HAVE A STAFF CAFETERIA? Do you eat there on a regular basis? If so, have you ever taken a close look at it? It's worth doing so since a staff cafeteria reveals a lot about your employer, for example, how much they truly care about your health. While some cafés are a haven for followers of the paleo (or other protein-based) diet and serve up steak three times a week, others cater mainly to vegetarians. And some actually still display 'sample plates' featuring shriveled sausages or pallid fish, which should really no longer be acceptable in this day and age when the aestheticising of food has become the norm.

Offering food options that are so unappealing that they trigger your flight reflex rather than your appetite would be unthinkable at a place like Google.

No other cafeteria in the world is as publicly celebrated as theirs. The word *cafeteria* alone is an insult. Google's offices boast a number of well-designed restaurants, bistros, cafés, and microkitchens, which cater to pretty much every culinary desire. The company is famous not only for its data-collecting mania but also for its monitoring of its employees. Google's thinking is that healthy employees (i.e., employees on a prime diet) are happy, and happy employees do outstanding work. They are innovative and creative. Therefore, at Google, all food is free, whether it be sushi, Thai cuisine, or Indian—you name it—and is available to everyone 24/7, lest they consider leaving the campus. If they fancy a snack, staff merely have to walk a couple of metres to one of their microkitchens stocked with a host of (healthy) choices.

There is a downside to free food: It is tempting even when you're not hungry. Except, Google doesn't want its employees to overeat, and so uses certain psychological tricks. The food cosmos is a cleverly designed arena of manipulation. As soon as you enter the café, the salad bar catches your eye—people tend to help themselves to what they see first. Sweets such as M&M's are kept in opaque containers, which curbs employees' calorie intake. Google also encourages employees to reach for a small plate rather than a large one. Chef Scott Giambastiani put Google's food philosophy this way: 'We make it convenient for people to eat healthy. We're quietly nudging you with products like organic fruit and greens, and dried fruits.' Google wants to create the healthiest workforce on the planet. And people eat less when they use small plates, leading them to feel less tired later in the day. There's a direct correlation between food intake and productivity, and that's worth it for Google. But since smaller plates don't necessarily mean that hungry Googlers will make healthy choices, a colour-coded traffic-light system helps them make their decisions. A green dot on a food item signifies: 'Eat me any time.' Yellow dots mean: 'Eat me only sometimes,' and red means: 'Not too often please!'

Gently directing people's behaviour in this way is called *nudging*. In their book *Nudge: Improving Decisions About Health, Wealth, and Happiness*, Richard Thaler and Cass Sunstein write that nudging is also at play when a mirror is hung behind a buffet: People help themselves to more fruit and fewer donuts. When confronted with an image of our not-so-healthy selves, we're compelled to reach instead for an apple.

A study in the Netherlands in 2015 examined whether nudging is as effective when it's transparent. Researchers conducted a field experiment at three snack-laden kiosks on station platforms, a setting where people make impulse purchases. At the first kiosk, unhealthy snacks—chocolates and biscuits—by the cash register were replaced with fruit and muesli bars. The second kiosk was left unchanged, and in the third the healthy alternatives were also positioned by the cash register but this time with a note reading, 'We help you make healthier choices.' The result: Where the fruit was positioned by the cash register, weekly sales of these items increased significantly, with or without the 'nudge' note: 287 and 245 healthy snacks sold at the first and third kiosks, respectively, versus 161 units at the second kiosk, which was un-nudged. Apparently, and here's the interesting bit, consumers didn't feel patronised by the note. In fact, many were glad to have the healthy choice so easily at hand! Sales of unhealthy snacks, however, were unaffected. Were people compelled to buy a healthy snack in addition to an unhealthy one? Were already healthy people nudged into buying snacks?

FOOD FOR THOUGHT: The Google food program is a central pillar of the all-pervasive optimisation culture the company's employees are required to embrace from day one. The question remains: What happens to those who despite all the nudging aren't deterred by red dots and make the wrong choices anyway? Does Google count calories in secret? Do they sentence the sinners to a training course in diet and nutrition? Does an alarm go off every time someone grabs a can of Coke? An optimised food world has no provision for deviations from the plan. But what about your own employer? Do they nudge you in a healthy direction? What other forces are nudging you to make healthy—or unhealthy—food choices?

CHAPTER 29

The Food Radius

How can you shape your food environment—and how does it shape you?

No sooner have we gotten out of bed in the morning than our thoughts turn to food—and not only when last night's steak is sitting heavy in our stomach, putting us off breakfast. Even on average days there are many questions to answer: Who's making coffee, and what are we having with it? One croissant or two? What to pack for lunch? Where, and more importantly, with whom are we going to spend our lunch break? Sushi or pizza after work? And who's cooking this weekend? After all, it's Thursday already. What's missing in the pantry, and where shall we buy the best supplies for the upcoming BBQ? Should we try that new delivery service?

It may come as a surprise to hear that in these times of the flexible, modern-day human, we still make eight out of ten food choices at home and within our own neighbourhood. Our immediate living environment and a short radius of less than seven miles dictate how and where we satisfy our hunger and culinary cravings on a daily basis. The psychologist Brian Wansink calls this our *food radius*. Not many of us realise the impact the nature of our personal food radius has on how and what we eat. How could we? After all, we are the ones who actually make the decisions—or at least that's what we believe. It is usually not until our food radius changes, for example, when we're on holiday, that we realise just how ingrained our food habits have become. The minute we arrive in Paris, we are already looking for our favourite fast-food restaurant near the hotel or checking if the supermarket around the corner stocks decent beer. It doesn't matter where we are, we immediately check the territory for its culinary suitability and establish rituals. Most of us are extremely ritualised—or lazy—with regard to shopping and eating.

This laziness follows a principle in ecology—the study of how organisms relate to their environment—called *optimal foraging*, which suggests that we prefer food sources that provide maximum energy output but require minimum energy input. In other words, we want to satisfy our hunger and food preferences without travelling too far. There is usually no lack of easily obtainable foodstuffs in our food radius—unless we live in the desert. On the one hand, it's comforting to know that food is always within reach; on the other, we find ourselves permanently surrounded by temptation.

The fact that your food radius covers the same geographical area as your neighbours' does not mean that you always run into them while shopping for groceries. Everyone uses their food radius differently. There are presumably dozens of restaurants and shops in your neighbourhood that you have never even heard of! Yet you barely visit more than a handful of them.

Can we influence our food radius, and can we shape it? Of course! A citizens' initiative in Berlin did just that. Instead of allowing yet another chain supermarket to be built under their noses, they developed plans for weekly markets geared towards specialist food traders, regional products, and maximum variety. The project gave birth to Markthalle Neun (Market Hall Nine), a restored nineteenth-century covered market that has become famous far beyond the city gates. Dozens of vendors offer affordable food of all kinds—vegan, paleo, Bavarian, Peruvian, organic, or simple traditional fare.

Sometimes all it takes is a focused walk around your neighbourhood to expand your horizons and discover new food sources in the vicinity. It's astonishing how much our perception changes once we resolve to do something. You may have had an experience like this: You want to buy a new bicycle, and you spot a bicycle shop on every corner; you are planning on having a baby, and you see expectant mothers everywhere. If you were to resolve to buy more locally grown food, you would probably discover farmers, community gardens, or markets offering fresh regional products not far from your home.

FOOD FOR THOUGHT: If you find gardening too boring and are not quick to develop neighbourly affinity, imagine you have recently moved to the area and rediscover your food radius—with an open mind, of course. You will be surprised at the possibilities on offer.

So if you want to better understand why you eat *how* you eat, you should examine your food radius carefully. Which shops reinforce healthy habits and which encourage unhealthy ones? Why do you frequent the ones that you do? Maybe because you enjoy chatting with the owner of a particular shop? Do you often buy more from them than was on your shopping list out of sympathy? Do you grab a bag of frozen vegetables at the supermarket because it is more convenient than taking a detour via the organic produce shop? How about setting up a syndicate with your neighbours and doing the vegetable shopping in turns? Or why not go a step further and share a communal garden with your friends?

CHAPTER 30

The Trophy-Kitchen Syndrome

Why doesn't renovating your kitchen always make you happy?

One of the distinguishing features of the so-called trophy kitchen is that it is big enough to roller-skate around or to play hide-and-seek in its spacious pantries. Trophy kitchens offer open dining, in which, because of their size, the waft of lingering cooking scents is never a problem and won't ever reach as far as the table. It goes without saying that they are built only of the finest materials. According to *The Sage Encyclopedia of Food Issues*, items commonly found in a trophy kitchen are countertops made of granite, marble, or quartz, lots of stainless steel, and professional kitchen equipment, plus special gadgets such as built-in espresso machines and warming drawers. Kitchen islands and stagelike lighting that draw attention to the high-tech equipment in the glass cabinets are also an integral part of such a kitchen. As the name suggests, the point of a trophy kitchen is not only to be a place for cooking but to be a status symbol—a well-designed centrepiece that reflects its owner's lifestyle and personality.

It is no coincidence that popular culture has immortalised the trophy kitchen with one example after another. The director Nancy Meyers, whose film sets look like something from an interior-design magazine, featured a shiny trophy kitchen in her successful 2003 film *Something's Gotta Give*. Its white cabinets and soapstone-like benches looked so pure and cozy that they made you want to move in immediately. In the film, Diane Keaton and Jack Nicholson spend so much time cooking, talking, flirting, and just hanging out in their model kitchen that Meyers has become something of an icon in the world of glamour kitchens. *Architectural Digest* devoted several pages to her and her work.

We covet these trophy kitchens, but often, once we have them, we still aren't happy. *The New York Times* has focused on the phenomenon and even identified this condition as 'post-renovation depression'. Once the remodeling work is done and the new kitchen is finished, the passionate amateur designer falls into a deep mental hole, a bit like becoming an empty nester just after the children leave home. However, post-renovation depression is not the only downside of our interior-design perfectionism: Believe it or not, a luxury kitchen increases the likelihood of snacking. The happier we are in our kitchen, the more time we spend in it, and the more often we open the fridge and help ourselves to what we see. We've all been there. Here, Brian Wansink comes back into play. He sees kitchens like the one in *Something's Gotta Give* as the manifestation of a society that endorses mindless eating. Wansink is not only a popular speaker; he also does kitchen makeovers. His philosophy is that it isn't enough to try to be rational about food; you need to outsmart yourself. Should you happen to have a sofa in your kitchen, Wansink would advise you to remove it. The same applies to a TV. Homey lighting? Bad. A charger for your iPad? Bad. Designer chairs that don't only look good but are comfortable, too? Also bad. Is there sensory overload? Are crisps, Nutella, cornflakes, and biscuits on display while fruit and vegetables are hidden away? Very bad indeed. An empty kitchen is no solution either, as you're more likely to go out for a meal when you're hungry. Wansink advises against white or cream-coloured walls as they stimulate the appetite.

Instead he suggests blue, green, gold, or earthy tones. Pumpkin also works well, he says—the colour of his own kitchen.

FOOD FOR THOUGHT: It is quite possible that if you follow Wansink's advice and remodel your kitchen accordingly—if you repaint it, install harsh lighting, and replace unhealthy snacks with carrots, kale, and cucumber—you will spend less time in the kitchen and cook more healthily. But at the same time it would rob your kitchen of its social complexity. It would no longer be the heart of the home, no longer a place for casual get-togethers, familial communication, or romantic (or even erotic) moments. The kitchen would diminish into a dull and joyless room where not even the act of cooking could kindle the flame of passion. And who would want that? It's natural to lust after a trophy kitchen, but remember that it won't necessarily improve your life—or eating habits.

CHAPTER 31

Working Lunch

For business, for pleasure, or for health?

ONCE UPON A TIME IT WASN'T uncommon for business men and women to allow several hours for a lunch with clients or even to clear the calendar completely for the whole afternoon. They would tuck into their food with gusto, drink (probably martinis; definitely alcohol), smoke (cigars), and generally enjoy themselves. Detailed agendas to drive the conversation towards a formulated goal were unheard of. Discussing business before dessert was taboo. And it wasn't even that long ago. This was the sixties and seventies—a time that the Swiss entrepreneur Philippe Stern, former CEO of the Geneva-based luxury watch manufacturers Patek Philippe, remembers well. In a nostalgic interview, he revealed, 'You would show your appreciation by taking your clients to the best restaurant in town. In the morning you would discuss business matters, and then you'd go for an extended lunch, maybe until half past two in the afternoon.' When Stern's father was still involved in the company, he would sometimes invite clients to his home. Aperitifs in the garden would be followed by lunch and then a boat ride on Lake Geneva. 'At the time, our company headquarters used to be by the lake in the city centre, and my father had his moorings right outside the office. He'd land there with the clients at four in the afternoon. Today that would be considered wholly improper behaviour.'

In the twenty-first century, a business lunch follows the dictates of efficiency and economic constraints. It's not designed to be social entertainment; after all, the aim of a business meal is, as the term suggests, to kill two birds with one stone. As the business lunch became more functional, the food itself followed suit. You shouldn't order food that you fancy but instead what is most practical to eat—on this the professional etiquette guides all agree. More specifically this means: Don't order a dish you can't pronounce properly. Avoid food that requires too much of your attention, that is likely to get stuck between your teeth, or that can't be eaten quietly. Opt for grilled fish, grilled meats, or salad (but not a tiny salad; you don't want to give them the impression that you're a picky eater). If you're on business abroad, eat what you're given, even if you have no idea what's been put in front of you. In short: The less fuss you make, the better.

Some even go so far as to ridicule the idea of lunch altogether; as the ruthless stockbroker Gordon Gekko famously said in the 1987 movie *Wall Street*, 'Lunch is for wimps.' That is, with the exception of so-called power lunching. In a book with that title, published in the 1980s, the authors Ligita Dienhart and E. Melvin Pinsel explain that a business lunch is to a power lunch what a lightning flash is to a laser beam. A power lunch is more serious, more focused, more powerful. But it's not only the business lunch that has suffered a loss in significance; lunch itself has become much less important, too. Many managers these days prefer their employees to eat their lunch (preferably something healthy like crudités, quinoa salad, or a whole-grain sandwich with hummus) at their desks. Multitasking instead of taking a break.

Your manager may think that working through lunch is productive, while in fact it's quite the opposite. In an interview with *Time* magazine, management professor Kimberly Elsbach has found, 'Never taking a break from very careful thought work actually reduces your ability to be creative. It sort of exhausts your cognitive capacity and you're not able to make the creative connections you can if your brain is more rested.' Taking a lunch break helps you be *more* productive, not less. 'If you're skipping lunch to continue to push forward in a very intense cognitive capacity, then you're probably not doing yourself any favors.' A break sometimes provides needed perspective on a problem and is a great opportunity to socialise with coworkers or step outside for a few minutes. Research has found that people who take a walk during lunch experience a significantly better mood for the rest of the day.

Given that we live in a knowledge-based society, it is all the more astonishing that we ignore so much of what we know about healthy eating and instead live constantly on the go. When the Mexican film director Alejandro González Iñárritu was asked by a reporter at *The Wall Street Journal* to identify the biggest culture shock he experienced after moving to the United States, Iñárritu replied, 'Eating from plastic in offices. I couldn't understand it. . . . [People] would have this food come in, and they'd eat it at their desks in plastic containers using plastic tools. It was shocking to me.' He added that he missed what they call *sobremesa* in Mexico, the time spent sitting at the table after a meal, in no hurry: 'It's when you have a little cigarette with your wine and then—deliciously gossip. It's that pleasurable moment.'

FOOD FOR THOUGHT: Lunch is for winners. Common sense alone dictates that a break from work (for at least thirty minutes) doesn't harm but rather boosts productivity and that an hour spent away from annoying colleagues has great potential to recharge one's batteries. Countless studies have shown the importance of breaks for creativity and recovery—spent alone or with friendly colleagues, and preferably including a little walk. So, then, is the new trend of skipping lunch altogether truly maximising productivity or reducing it? We may not have a boat on Lake Geneva at our doorstep, but in what other ways can we add value to a lunch break?

CHAPTER 32

Fast Manners

How did handling fast food get so out of hand?

Picture this scene, it could be anywhere: A middle-aged man in a suit is sitting in a packed bus, eating a bacon, lettuce, and tomato sandwich. Every time he takes a bite, a blob of mayonnaise escapes down the side, and so every now and then he unashamedly licks his fingers as if it were the most natural thing in the world. This person without manners could of course be a woman, too—everywhere we look, people are eating, slurping, chewing, or drinking, be it at the hot dog stand, on the bus, on the train; while walking, shopping, or riding a bike. Unconcerned, we sate our hunger as if everywhere was McDonald's. Rejecting table manners and devouring burgers from greasy paper boxes in a fast-food restaurant used to be a way for teenagers to distance themselves from the world of adults, that is until slowly but surely people of all ages came to love this fast-food and snacking culture. Today, going out socially always involves eating something; it's impossible to imagine one without the other. A date isn't a proper date unless it involves plenty of food. Going to see a movie would be half as much fun without popcorn, according to the ethnographer Phillip Vannini. A day on the beach without ice cream—unthinkable. And no picnic would be complete without a watermelon. We have become incidental eaters. Nonetheless: We are not only *what* we eat, but also *how* we eat.

Somewhat pessimistically, the cultural scientist Walter Leimgruber predicts not only the demise of the use of cutlery, an important discipline in the civilising process, but even its reversal. 'How else,' he asks, referring to the sociologist Norbert Elias, 'can it be explained that all of his rules, table manners, the custom of eating with a knife and fork, are increasingly on the decline and are giving way to unchecked food consumption anywhere and everywhere? Often, we are oblivious to the actual smell or the flavour of what we eat. Those ketchup-and-mayo guys who drown everything in condiments are a direct consequence of food products that have no taste at all.' We know only too well where such mindless consumption is taking us (and in the United States, where it has taken us already): to excessive food intake and obesity.

Fast food is no modern invention—the ancient Romans and Greeks would snack on fried fish, salted peas, or bread—but it was industrialisation that laid the cornerstone for today's takeout and snacking culture. In England, thousands of fish-and-chip shops sprung up in the vicinity of factories and workers' housing estates, while Germany's popular sausage stalls didn't appear until after World War II. Ultimately, the 1960s heralded the triumph of fast food for good: It started with Kentucky Fried Chicken, followed by McDonald's, then Pizza Hut, and so on.

Our love of the burger especially seems unabated, although the trend has been moving towards better-quality, creatively stacked 'premium' burgers. In contrast to the large chain restaurants that still push the old run-of-the-mill burger, chic new burger bars have remodeled the 'premium' burger into some kind of delicacy by adding variable layers of goat cheese, avocado, kimchi, or caramelised onions. Large cities have seen one such shop opening after another. These days even sushi is classified as fast food, although raw fish would never be labeled as such.

FOOD FOR THOUGHT: There's a strong correlation between the rise of fast food and the decline of table manners. What we're gaining in convenience, we're sacrificing in health and quality—both of the food and of the time we spend eating it. Even if you do occasionally go for fast food (we all have those days), try sitting down at a table and picking up a knife and fork instead of using your hands. Fast food may be more popular than ever, but that doesn't mean we'll all turn into 'those ketchup-and-mayo guys'. And there's hope for taste, too: As healthy eating and the joy of food are gaining in importance once again, the limp and tasteless burger at least is losing its appeal. Sitting at a table helps you eat more slowly (and ultimately consume fewer calories), and gives you more satisfaction from your meal. Not to mention, you'll be more attractive without ketchup on your shirt.

CHAPTER 33

Stress-Free Slurping

Why is drinking a milkshake so calming?

IMAGINE YOU'RE STUCK IN TRAFFIC. It's oppressively hot, the air-conditioning is broken, and a fly is buzzing persistently around your head while your car hasn't moved an inch. How do you deal with stress in a situation like this? Let's hope you wouldn't go Michael Douglas' route in the movie *Falling Down* and abandon your car in the middle of the freeway out of frustration, then walk across LA armed and angry. If only he'd had a milkshake.

The economist and Harvard professor Clayton Christensen led a study on behalf of a fast-food chain to find out which marketing measures would be the most effective to increase its milkshake sales. The researchers started by identifying the core demographic; in other words, they determined who were the people who would normally buy a milkshake. Maybe you're thinking of children, or teenagers, or moms-to-be. Far from it. Christensen found that almost half of the milkshakes at that establishment were purchased before 8:00 AM. Is this the case for the entire country? While most children and teenagers are still at home having their breakfast, classic milkshake buyers are already on the move. Who are these consumers? That would be commuters facing a long journey to work.

So then, 'What job are customers "hiring" the milkshake to do?' asked Christensen. Or, in layman's terms, what are customers buying the milkshake for? It's an interesting turn of questioning that focuses on social dimensions rather than on the product itself. After all, it would make no economic sense to offer twenty different milkshake flavours when your customers don't have the leisure to make time-consuming decisions. It became apparent that 'they all had the same job to do in the morning: They had a long and boring drive to work and they just needed something to do while they drove to keep the commute interesting,' said Christensen. 'They weren't hungry yet, but they knew they'd be hungry by ten o'clock.' In short, sucking a thick liquid through a thin straw gave them something to do on a monotonous commute to stave off hunger and boredom.

Another simple but often neglected feature of a milkshake is the calming factor of the straw. It's gratifying not only because the straw stops the liquid from dripping onto our work clothes, but also because the sucking action has a soothing effect. To start, it changes the consistency of the milkshake in the mouth, where a small amount of liquid meets with a larger amount of air. The result is a pleasantly creamy sensation. Because of this, the liquid stays in the mouth for longer, intensifying the flavour—especially its sweet notes. The act of sucking is also associated with pleasure, reassurance, and satiation: It is one of our earliest experiences of all, an instinctive, vital reflex. As we grow older, the reflex is superseded by a habit accompanied by positive feelings. And it is this association that drives us to buy a milkshake. That craving is controlled by the hormone dopamine. Our brain releases dopamine when we are looking forward to something, which is why it is sometimes called the 'I want' hormone.

Could a chocolate croissant do the job of a milkshake? No. It's eaten too quickly, it's too flaky, it makes a mess on the seat, and it isn't filling. Instead of calming you down, it creates even more stress. And commuting is stressful, even if commuters themselves can't always see the effects of that stress. It adversely affects your sleep, increases your heart rate, builds tension—even more so when you can't reliably predict what time you'll reach your destination. Not to mention the social consequences. Over a period of ten years, a researcher from Sweden gathered vast amounts of data, which showed that commuting increases not only the likelihood of earlier death but also the divorce rate. For couples in which at least one of the partners has a commute of forty-five minutes or more, the risk of separation increases by a whopping 40 per cent.

FOOD FOR THOUGHT: We now know that the sensation of drinking a milkshake is much more important to our experience than its taste. The problem is that most of the ones we pick up on our way to work are laden with unnecessary amounts of sugar and saturated fats, often amounting to more than 1,000 calories in a 700-millilitre cup. For a healthier alternative, try making them at home by blending your favourite fruit with half a cup of milk or yoghurt (120 millilitres of milk; 120 grams of yoghurt)—and don't forget to keep some straws on hand!

CHAPTER 34

The Comfort-Food Effect

Why do you crave junk food when you're sad?

Experience, literature, film, and hearsay all teach us that emotions can drive a person insane. What may be new, however, is the scientific finding that the physical experience of romantic rejection is similar to a junkie's excruciating desire for a fix. The American psychologist Arthur Aron, whose research focuses on the complex architecture of interpersonal relationships, has examined the brain activity of heartbroken individuals. Merely looking at the photograph of a loved one was enough to trigger physical withdrawal symptoms usually seen in addicts. Bridget Jones, the protagonist from the eponymous novels and films, knows this condition only too well.

Hearing a certain song on the radio makes her collapse in a sobbing heap on the floor. To make herself feel better she turns to ice cream and chocolate. In a state of acute sorrow, these foods were her indispensable source of consolation, her comfort food—food that nourishes mainly one thing: emotions.

But what exactly is it that triggers these intense cravings for ice cream, chips, and chocolate in particular moments? Apart from the ingredients—the combination of carbohydrates and fat kicks off various hormonal regulating processes that lead to the release of the 'reward hormone' dopamine—the answer lies in our senses. We have a particular fondness for foods that melt in the mouth. Why? Because dealing with the changing consistency in our mouth requires our attention. It allows our inner self to relax for a moment. For at least as long as we concentrate on the pleasure of eating, we are distracted. It's this state of distraction that motivates us to go off in search of our favourite comfort food as soon as an emergency arises.

How receptive we are to comfort foods may also be a question of type. Psychologists Jordan D. Troisi, Shira Gabriel, and others discovered that the people who prefer comfort foods tend to be securely attached (meaning, in psychology, that they form healthy relationships). Emotional stress makes these people want to retreat back to the kitchen table of their childhood. It's no wonder that favourite dishes from our childhood days such as pancakes, mac 'n' cheese, and a hearty chicken soup are popular comforters for our soul. The mere smell of your favourite food can provide comfort. Yet the things that really lift our mood are loving care, a smile, and encouragement—and the scripts behind good TV commercials draw on exactly these needs. Emotional images tend to connect particularly well: people showing love and kindness, strangers smiling at each other, a playful family enjoying their Sunday breakfast around the table. Every ad tells a story. The more credible the narrative, the stronger the emotional connection to the advertised product that we're supposed to reach for every time we long for a certain feeling.

The soothing or stimulating effect of the ingredients plays directly into the hands of the manufacturers. Cocoa beans, for example, contain the essential amino acid tryptophan, precursor to the happiness hormone serotonin, and the caffeine-like theobromine. You can have too much of a good thing, though: A constant hyperactivation of serotonin release can use up the limited supply of neurotransmitters in the brain, eventually leading to habituation instead of the desired effect of feeling good.

From emotional stress to inner peace thanks to comfort food? Sadly not. Reaching for a candy bar every time you're feeling low isn't always the best choice. In another experiment, this one by researchers at the University of Minnesota, participants were shown a compilation of sad and upsetting scenes from movies and then given either their favourite comfort food, a snack they liked (such as a muesli bar), or nothing at all. The result was always the same: Their mood improved over time, with or without comfort food, even though more than eight out of ten participants had previously claimed that they believed their go-to anti-stress food always made them feel better. Seeing causality in the consumed food is a faulty association, study leader Traci Mann said. Comfort food is a myth. Although you may be looking for a justification to eat high-calorie foods such as ice cream, it doesn't have any magical mood-boosting abilities. Nonetheless she recommended, 'Just eat the ice cream! It's not magical. But it is yummy.' The advice doesn't apply to emotional eaters who automatically counter negative emotions such as stress, loneliness, or boredom with salt and sugar. Anyone who has lost control of their eating habits is caught in a dangerous downward spiral and may require professional help from a therapist to get back on track.

It is undeniable that emotions affect our judgement. When we are agitated we find it much harder to gauge a food's calorie or fat content. Participants in an experiment were shown one of three videos deemed either happy, sad, or neutral. Afterwards they were asked to estimate the amount of fat contained in samples of milk mixed with heavy cream in varying amounts. Those who had seen the sad video significantly underestimated the fat content. This effect was not observed among those who had watched the neutral video. It goes to show that our working memory has limited capacity. Losing oneself in an emotional situation and at the same time correctly gauging the fat content of a plate of french fries really isn't doable.

FOOD FOR THOUGHT: The idea that comfort food will make us feel better has been ingrained over a lifetime. But is it really that favourite food that lifts our mood? Maybe not, but indulging in that high-calorie snack still releases dopamine and stimulates pleasure. We're all addicts when it comes to chocolate. That spike of dopamine is a welcome high when we're sad or depressed. But what about when we're happy? Why reach for the chocolate bar then? Chocolate may be good for the soul, but it is not good for the waistline.

At least we now have a scientific explanation for why a broken heart makes us reach for the chocolate bar. And knowing that our mood will improve even without the candy is a boon, and not only for our waistlines.

PART VI

How You Choose with Your Tongue

CHAPTER 35

Supertasters

Do children who will eat only pasta have supertasting powers?

You have probably wondered why people sharing a meal together experience and rate identical dishes so differently. Some people add copious amounts of salt while others grimace because the artichokes are too bitter or the strawberry parfait is too sweet. 'Tastes differ,' we are told from an early age—without being given a reason. And if we were, that reason would probably have been wrong. The tongue was long understood to be some kind of map with clearly distinct regions for the different elements of taste perception: sweet, sour, salty, bitter, and (later) umami. In reality, however, each taste can be detected anywhere on the tongue. There may be dominant regions, but taste buds, sitting on visibly raised papillae, are located all over the tongue. Each individual taste bud contains up to 150 receptor cells that distinguish the various gustatory qualities of our food.

This does not, however, explain why some people are more sensitive than others to a particular flavour. It is partly due to the number of taste buds a person has, as taste researcher Linda Bartoshuk at the University of Florida Center for Smell and Taste discovered in 1993. She dyed the tongues of her research participants blue and counted the protruding papillae. The number and size of the papillae differed considerably among the participants. In a taste test, those with a particularly large number of small papillae turned out to be rather sensitive. Participants were asked to rate the bitterness of the compound propylthiouracil, or PROP for short. Those with many small papillae rated it as extremely bitter, while the other participants rated it as neutral or at most somewhat bitter. Bartoshuk called the first group *supertasters*. People who experience taste with an extreme intensity live in a 'neon-lit' taste world, while the rest of us live in a 'pastel world', she explained.

The geneticists Michael C. Campbell and Sarah A. Tishkoff examined the differences in sensitivity towards bitter tastes among African populations and discovered an astonishing number of genetic variations of bitter receptors. The fact that supertasting abilities are determined by our genes explains why taste sensitivity is often inherited. If one of your children happens to be a picky eater and won't eat anything but pasta, that child may have inherited a supertaster gene (even more likely if there are other picky eaters in the family). Just as gender and eye colour vary between siblings, so does the makeup of the tongue. Supertasters often find apples too sour, pickles too salty, grapefruit too bitter, and caramel candies too sweet, as their intense sensitivity to taste is not limited to bitter compounds.

Nevertheless, these foods should be given their rightful chance (except maybe the caramel candies). After all, a picky eater may eventually learn to enjoy common tastes thanks to a special learning method: habituation. When a body is repeatedly presented with a stimulus, it gets used to that stimulus and will eventually cease to respond to it. While a first bite of a food to which you are sensitive elicits a strong, sometimes extreme response, as few as three or four bites later the reaction is distinctly less intense. That's one reason why molecular gastronomy, a cutting-edge cuisine, focuses on small portions and great variety.

If you have had no problems eating Camembert, endives, and pretzels from an early age you probably don't belong to the 25 per cent of people who are supertasters. You are more likely to fall into the category of medium tasters along with half the population, or you might even be a nontaster (the other 25 per cent), whose taste is distinctly less differentiated. Nontasters are extremely eager to try new foods. You may hear them say, 'I want to eat something I've never eaten before.' However, the opposite is true for supertasters—that is, unless they've overcome their aversions and turned their superpower into a fitting profession such as chef or restaurant critic.

FOOD FOR THOUGHT: Don't be too hard on your child if all they want to eat is pasta with ketchup. They may just happen to have a very large number of papillae on their tongue—and are more discerning than you are!

Are there any foods you simply can't stand? Have you tried them out recently? It's possible you've become habituated over time and have since acclimatised to that once-reviled vegetable. But if your mom or dad hates it, too, you may just be out of luck.

CHAPTER 36

Some Like It Hot

What does a fondness for spicy foods reveal about your character?

IMAGINE YOU ARE ON A DATE at an Italian restaurant. Like all people who harbour affection for each other, you both delight in discovering the things you have in common, such as a preference for the same wine or a passion for a certain movie. But suddenly things take a turn for the worse. Your date pours half a bottle of Tabasco sauce onto his pizza arrabbiata, offers you a slice drenched in the sauce, and goes on to enthuse about the various spicy dishes at his favourite Mexican restaurant, which you 'must try soon'. Until now you had actually been quite happy with the choice of restaurant, and the spaghetti carbonara on your plate is just how you like it. You were about to order your favourite dessert, panna cotta, but the prospect of sharing spicy foods with this guy in the future has put you off.

What if your unease at discovering these opposing food preferences was not unfounded? 'Because what I eat, what I drink is in itself the "second self" of my being,' wrote the philosopher Ludwig Feuerbach. What does that mean for how you see your date? Chilli likers are adventurers. This conclusion came from Paul Rozin and Deborah Schiller of the University of Pennsylvania after they evaluated the first systematic study on chilli ingestion. In Mexico, for example, eating chilli peppers is regarded as a sign of strength, valor, and masculine attributes. It was found that American students with a penchant for chillis also had a passion for daring and potentially risky activities including fast driving, parachuting, or swimming in ice. Each of these experiences initially requires a certain mental effort to conquer one's fears, but just as with eating chilli peppers, one learns to assess the risk over time. The constrained risk might just be what makes chillis so exciting for some, says Rozin.

Technically speaking, pungency is not a taste; it's not sweet, salty, bitter, sour, or umami. Hot means pain, which is why you, the careful type, automatically recoil from the offered pizza slice. The typical pain response is triggered as capsaicin, a chemical compound found in chilli peppers, hits the pain receptors on the tongue. If it got into your eyes or touched the sensitive lining of your nose, as capsaicin-containing pepper spray would, you'd be doubling over and screaming in pain. So, what makes someone abuse a sensitive part of their body like their tongue with a chemical weapon and reach out for products with names such as Spontaneous Combustion, the Reaper, Mega Death Sauce (Feel Alive!), and Pain 85%, 95%, and 100%? These may sound like the names of death metal bands, but they're actual off-the-shelf hot sauces.

Some people do it for pleasure. According to Rozin, inducing this negative physical experience (accelerated heart rate, sweating, burning sensation, watering eyes, shortness of breath) is evidence of benign masochism. He compares it to watching a horror movie that makes you feel real fear despite knowing that nothing can happen to you. You won't find anything comparable in the animal kingdom. Even pigs, who usually devour anything edible (and who, if they live in the Mexican highlands, should be accustomed to spicy leftover food from people), tend to give a wide berth to tortillas with a spicy sauce. They have no idea that the perceived heat isn't 'real' but merely a misinterpretation by their brain. We, however, know that chillis won't burn us on the inside. We are intelligent enough to defy the warning signals to a certain extent. In his book *Pain: A Story of*

Liberation, published in German, the physician Harro Albrecht writes that it is as if we triumph over a basic instinct from a safe distance and are rewarded by our brain with a biochemical treat in the form of endorphins. The same principle applies to the so-called runner's high of marathon runners.

Chilli lovers are eager to try new things; willing to take risks; and hungry for variety, strong emotions, and adventures—all characteristics linked to thrill seekers. Put in a good light, this means they have a high degree of curiosity; in less friendly terms, they are easily bored. This knowledge about your date's personality should send alarm bells ringing if you're someone who likes to play it safe. People who avoid excitement, value consistency, and generally manage perfectly well without exposing themselves to extreme situations should pay more attention to a new acquaintance's preference for hot spices in the future—even if chillis are beneficial for your health, boost your metabolism, and have an analgesic and antibacterial effect.

So far, so good, but the relationship between taste and personality doesn't stop at spicy food. What might having a sweet tooth say about you? Your date can count themselves incredibly lucky to have you by their side. People with a sweet tooth are regarded as extremely helpful and social (real 'sweethearts').

Experiments have shown that people who like to snack on chocolate have a more pronounced willingness to help others in need than those who opt for a salty pretzel. Sweet foods also relieve the symptoms of chilli ingestion, especially dairy foods such as panna cotta, mascarpone, or creamy desserts whose high fat contents bind the capsaicin on a molecular level. Tap water is unhelpful in putting the fire out as it merely spreads the capsaicin around and amplifies the burning sensation.

FOOD FOR THOUGHT: You may be averse to the sensation of spicy foods, but remember the thrill that chases the heat! A penchant for chillis may spice up your relationship, but make sure you know how to neutralise the heat—whether by keeping milk on hand or planning a relaxing evening for the next date. In dealing with a spicy person, know what approach is akin to water and what to milk.

You may have heard the one about people who add salt to their food without tasting it first (hint: don't do it). What might a fondness for salty foods say about a person? What about an inability to moderate when it comes to sweets?

CHAPTER 37

Conditions of Taste

Are your favourite foods innate or selected through memory?

ONE SPRING DAY IN 2011, the successful food writer Marlena Spieler left her San Francisco home to pick up a few local treats for her birthday party. As she was crossing the street, she was hit by a car. Spieler broke both of her arms and suffered a concussion, but as she wrote in a story about her accident in *The New York Times* in 2014, that was only the beginning of the nightmare:

> That night, acrid smoke woke me. Nobody was smoking and no one else around me seemed to notice it. My morning coffee was tasteless. Visitors brought delicacies to soothe me, but with each bite, I tasted dread. These well-chosen treats had no flavor, at least none that I recognized. Cinnamon drops, a childhood favorite, were

> bitter, horrible. Tamales were as bland as porridge. Bananas tasted like parsnips and smelled like nail polish remover. It got no better as days went by. Gently sautéed mushrooms seemed like scorched bits of sponge. Red wine was just flat and sweet or unremittingly bitter. I had lost my ability to taste and smell. It was like a musician losing her hearing.

In a flash Spieler's culinary archive had been erased. She notes that because her nerve was damaged but not severed, her memories of smell and taste could return, but there was no telling how long it might take. Once professionally trained in all things taste, she suddenly found herself in the role of the amateur again. It leads us to wonder, are we born with food preferences or are these learned through time and experience? Food-wise, there was no relying on anything. Food she never used to like now tasted delicious and vice versa. Every meal, everything she ate, existed completely detached from her life and her experience. This led to amazing taste experiences: 'Later, eating a bowl of ice cream, I murmured: "This is delicious. What is it?" My first mouthful of bacon was "So tasty!" But each encounter with bacon was like eating it for the first time. It might have been wonderful to be thrilled over and over again except that I felt incredibly stupid.'

The accident had cut the invisible thread that linked her to her culinary past. Her autobiographical memory—an important network of crisscrossing neural pathways—no longer performed its task as it should. Spieler had become a stranger to herself. No smell, no taste would suddenly resurrect any buried sensations or experiences. The mental journey through time described by Marcel Proust in his novel *In Search of Lost Time* had become impossible. It was the taste of a madeleine dipped in linden-blossom tea that delighted Proust's narrator with intense childhood memories: 'No sooner had the warm liquid mixed with the crumbs touched my palate than a shudder ran through me and I stopped, intent upon the extraordinary thing that was happening to me. An exquisite pleasure had invaded my senses, something isolated, detached, with no suggestion of its origin.'

Scientists studying human memory speak of the 'Proust phenomenon'. For some, the taste of creamy chicken soup brings to mind their late grandmother and they can imagine being there in the kitchen with her, while others are overcome by memories at the smell of roasted almonds. Scientists now know that memory is concentrated in several parts of the brain: The hippocampus plays an essential role in memory consolidation, while the amygdala is involved in storing emotional memory. Once a cue triggers a memory hidden far away in the depths of our mind, we are completely at its mercy. Smells imprinted deeply into our memory are among the most common and strongest cues for spontaneous memories.

Our taste memory goes back even further than most of us think: Conditioning begins in the womb, where the baby absorbs amniotic fluid, which has numerous flavours influenced by the dietary habits and food preferences of the mother. Before we can hear or see, we can taste and have our first olfactory experiences. About eight weeks into pregnancy, the taste buds are formed, and around the twelfth week the baby begins to swallow. During the last trimester, the baby's swallowing behaviour adapts to the taste of the amniotic fluid, of which it drinks half a litre a day: If the amniotic fluid tastes sweet, the baby swallows often; if it tastes bitter, the rate of swallowing drops.

A preference for sweet foods and an aversion to bitter substances is innate. When humans still lived in caves, this genetic programming ensured our survival. Sweet means that we are supplying our body with energy, while poisonous substances often taste bitter. The biologist Julie Mennella and her team at the Monell Chemical Senses Center in Philadelphia did an experiment with carrots to prove the great conditioning power of prenatal and early postnatal flavour experiences. They divided their pregnant study participants into three groups: The first group drank regular amounts of carrot juice during the last trimester of their pregnancy and switched to water during the first months of breastfeeding; the second group drank only water during pregnancy and started drinking carrot juice immediately after giving birth; and the third group didn't drink any carrot juice at all. When the babies were introduced to solid food, they were given cereal prepared with water during one session and cereal prepared with carrot juice in a second session. The results showed that the babies who were already familiar with the taste of carrot juice through amniotic fluid or breast milk ate more of the cereal made with carrot juice and expressed fewer negative emotions than the babies who were unfamiliar with carrot flavour. 'Each individual baby,' Mennella said in an interview on NPR, 'is having their own unique experience, and it's changing from hour to hour, from day to day, from month to month.' When a baby begins to eat solid food, he is likeliest to prefer, and to recognise as edible, exactly the same foods his mother is used to eating—from an evolutionary standpoint, familiar tastes present the safest option. Therefore, the healthier and the more varied a pregnant or breastfeeding mother's diet, the less fussy and more willing to try new foods her offspring will be. But it's not only carrots. Vanilla, garlic, aniseed, blue cheese, and mint are other strong flavours that find their way into a mother's milk.

Our favourite meal from childhood occupies such an important place in our autobiographical memory that it often remains our favourite meal for life. The fact that it will never again blow us away, even though it reliably comforts us, is the fly in the memory ointment. Or, as the Dutch novelist Cees Nooteboom so beautifully described it, 'Memory is like a dog that lies down where it pleases.'

FOOD FOR THOUGHT: Building a varied culinary archive is just as worthwhile as the search for lost time. We may have innate preferences from before birth, but those, too, were conditioned by exposure. Marlena Spieler tackled reprogramming her sense of taste head-on—with resulting ups and downs. We, too, can relearn taste! Don't be so quick to dismiss a once-loathed food based on memory alone. Consider what associations you have with that food, and why you may have come to dislike it. There's something magical about invoking one's past using sensory impressions. After all, sometimes the dog will lie down just where it pleases *us*.

Can we use taste to then store our memories? Next time you're trying to remember a conversation had over dinner, think about what you ate—or even try eating it again.

CHAPTER 38

Mind over Meat

Why do cats sit on your lap and cows on your plate?

It is often said that dogs look like their owners. Is this coincidence, a trick of nature, or just our imagination? 'The contemplation of animals delights us so much, principally because we see in them our own existence very much simplified,' wrote Arthur Schopenhauer, who was said to be quite fond of dogs. This may explain why some pets seem to be living a life of luxury that includes gourmet food and grooming and petting services. But livestock animals also warm our hearts. A lamb! How cute! A newborn calf, how pretty! The reality of what happens to go from a cute lamb in the meadow to roast leg of lamb on our plate is best kept hidden inside a black box. Our moral consciousness distinguishes clearly between animal and meat as if the one had nothing to do with the other. Even if we refuse to eat meat for 'ethical reasons', we don't necessarily think it incongruous to warm our feet in cozy lambskin slippers. In psychology, this is called *dissonance*. It occurs when a person has desires, expectations, and beliefs that are incompatible with their actions. People who praise their pets for their cleverness but at the same time deny livestock any intelligence of their own are caught in a dilemma. And so you must ask yourself: Is Skippy the dog (a) more intelligent than, (b) just as intelligent as, or (c) less intelligent than Bessy the cow?

Pigs are notoriously intelligent animals, something farmers know only too well, which is why they will always ensure that the latches on their gates are carefully secured. In his book *Eating Animals*, Jonathan Safran Foer recounts a story told by the British naturalist Gilbert White, who wrote in 1789 about a sow who was so artful she managed to open a gate and 'all the intervening gates, and march, by herself, up to a distant farm where [a male] was kept; and when her purpose was served would return home by the same means'. But who wants to hear stories about how clever animals are while they're tucking into a hot dog?

Australian scientists have studied this phenomenon in more detail. To begin with, students participating in their study were asked to assess various aspects of thirty-two different species of animals (both domestic and wild) and to indicate whether they ascribed various mental abilities, including fear, pleasure, morality, and memory, to the animal. Afterwards they had to choose: Eat it or leave it? The greater the intellectual capabilities they ascribed to the animal, the less keen they were to eat it. And we find this to be true every day. Chances are, you answered (a): Skippy the dog is more intelligent than Bessy the cow. The perceived intelligence of an animal affects its appeal as food.

Yet this doesn't go to answer the question of why we eat meat in the first place, especially meat that comes from animals we consider cute, intelligent, or capable of feelings. So the scientists did further research. This time, the participants were divided into two groups and given a questionnaire that required them to consult a picture of a sheep or cow grazing in a field and a description of its living conditions; however, the first group received the picture first and then the description, which indicated that the animal lived on a farm and had some freedom to move around, whereas the second group viewed the picture only after reading the description, which indicated that the animal was going to be transported to the abattoir, killed, and gutted, and its meat packaged for sale in the supermarkets. In the questionnaire, all participants were asked to assess the intelligence of the animal. The result was startling:

Those who were told before viewing the animal that it was raised only for meat considered it to be significantly less intelligent than those who saw the animal first and were told it was raised on a farm. Counter-intuitively, reading about the meat-manufacturing process reduced people's inhibition to eat the animal. To reduce the dissonance, they justified their decision by denying the animal any mental abilities.

The American psychologist Melanie Joy speaks of an invisible belief system that she calls *carnism*. The whole system is designed to block our consciousness and our empathy, and its success depends on our culture and the animal. Would you eat dog meat? No way, I've got a dog on my lap right now. Cat meat? Unthinkable; she warms my feet at night. Joy emphasizes that the mechanism of denial is very important in this judgement. Conveniently, denial is made easy for us. The animal arrives at the table as a processed piece of meat, described as a Wiener schnitzel, steak stroganoff, boeuf bourguignon, etc. According to Joy, further defence mechanisms include the 'three Ns of justification': that eating meat is normal, necessary, and natural. But sometimes empathy wins. When little Fern in *Charlotte's Web* saves the runt of a sow's litter from certain death and names him Wilbur, the piglet instantly turns from potential dinner to beloved friend.

FOOD FOR THOUGHT: 'Eating "man's best friend" is as taboo as a man eating his best friend,' writes Foer. Although eating dog is legal in almost every state, our inability to dissociate friend from food is so strong that most of us can't even fathom trying it. More and more we're developing empathy for farm animals, too—and the demand for ethically raised, 'grass-fed', and 'free-range' beef is on the rise. Not only is it better for our conscience, but the leaner beef is also better for our bodies. But doesn't cultivating empathy *discourage* us from eating animals? Once again, we're dissociating the life and intelligence of the animal itself from the actions we inflict in raising it for food. We rationalise by saying, 'Perhaps it's OK to eat *this* beef since it had a good life'—not because we recognise its intelligence.

CHAPTER 39

Hooray for Haptics

How do you experience the pleasure of taste through touch?

THERE WAS ONCE A CAFETERIA IN a large tech company that offered carrot salad several days a week. One employee complained about the carrots being sliced and suggested that they should instead be grated. The chefs replied that it's too expensive to grate the carrots and so they would remain sliced. Now, had the customer persisted, he might have been able to explain the importance of haptics for the enjoyment of food. Instead, he was ignored.

The way a food feels in the mouth influences how much we like it. Cornflakes with less crunch than a banana? Beer that doesn't foam; sparkling wine that doesn't fizz? Lukewarm chicken soup? Gross. If the mouthfeel is perfect, we are delighted. Think of the airy texture of a chocolate mousse, the fine vibrations of a sparkling mineral water, or a piece of meat that melts in your mouth, dissolving like butter.

Long before hearing and sight, touch is the first sense a foetus develops in the womb. It forms a neural basis upon which the whole nervous system is built. Free nerve endings in the outer skin layers register warmth, cold, stickiness, or the texture of food when touched by the lips, tongue, or fingers. We can be quite sensitive when dishes disappoint and downright euphoric when they exceed our taste expectations. But who actually says, 'It feels good in the mouth'? We don't usually venture beyond the words 'very tasty'. We rarely notice mouthfeel unless it differs from our expectations or is unpleasant, be it because we've burnt our tongue or because the carrot slices make the salad too awkward to eat. And yet we have numerous words that describe the haptics of the mouth: *silky*, *soft*, *tough*, *sticky*, *crispy*, *slimy*, *slippery*, *sharp*, *metallic*, *fizzy*, *hot*, *lukewarm*, *cold*, *grainy*, *granulated*, *melting*, *fiery*, *airy*, and so on.

Celebrity chefs such as Ferran Adrià and Heston Blumenthal like to mess with their customers' senses by stimulating mouthfeel. Blumenthal, for example, serves 'Hot and Iced Tea' at his restaurant the Fat Duck that is cold to the touch on one side and warm on the other. The customer, sipping the drink, is confused: Is he drinking hot tea or iced tea? Adrià used to serve grappa glasses filled with a green soup at elBulli (now closed). Customers who did as recommended by their waiter and tilted their heads back to drink the liquid would experience an amazing transformation as the hot pea soup was followed by warm and finally cold soup. Afterwards they'd be left with a strong minty taste in their mouths. Adrià changed the familiar form, texture, and temperature of food—to his customers' sheer satisfaction. And it goes on: What looked at first glance like couscous turned out to be cauliflower, separated into tiny florets the size of pinheads. Spaghetti carbonara arrived at the table as a translucent pile of light brown jellylike spaghetti, surrounded by a swirl of egg sauce, garnished with pancetta cubes and Parmesan, and drizzled with truffle oil. In the mouth, the chicken broth jelly melted and combined with the truffle oil to create the smoky flavour typical of a carbonara. Before it closed, this inventiveness had attracted up to two million reservation requests a year to elBulli.

Nonetheless, experiments with mouthfeel are nothing new. Almost a hundred years ago in Turin, Italy, the futurist Filippo Tommaso Marinetti challenged dining norms in his restaurant the Holy Palate. Dishes such as 'Pollock in Sunlight with Mars Sauce' and 'The Excited Pig' stimulated all the senses.

Guests were invited to 'a tactile dinner party' where they were asked to choose from among loungewear covered in different materials, such as cork, sandpaper, fur, small metal plates, silk, or velvet. Next, they would enter a dark room where they would carefully select a dinner partner by the feel of the person's suit. Together they would then be led to a table in the illuminated restaurant. They would place olives, fennel hearts, or kumquats into their mouths with their right hand, while their left hand would be stroking, say, the sandpaper on their partner's suit. The guests would also be served a 'tactile vegetable garden', which was to be approached in a prescribed way:

> A large plate containing a wide variety of raw and cooked vegetables without any dressing or sauce is placed in front of each guest. The greens can be nibbled at will but only by burying the face in the plate without the help of the hands, so as to inspire a true tasting with direct contact between the flavours and the textures of the green leaves on the skin of the cheeks and the lips. Every time the diners raise their heads from the plate to chew, the waiters spray their faces with perfumes of lavender and eau de Cologne.

But nothing riled the avant-gardist as much as pasta. He would have liked to abolish pasta altogether and replace it instead with foods free from 'volume and weight'. Cutlery? Banned from his table. Food was to be incorporated without interference. His specialty was 'magic food', served in small bowls dressed in tactile materials: balls made of burnt sugar, filled with candied fruit or garlic, banana purée, chocolate, pepper, or raw meat.

FOOD FOR THOUGHT: Not all touch-taste experiments turn out so pleasurable. Imagine eating ice cream, but instead of that smooth, silky texture you find little chips of ice, or biting into an apple and getting a mouthful of sand. Compared to that, sliced carrots from the canteen seems like *Haptics for Beginners*. Given how much we lust for haptic stimulation, it is surprising to see how often mouthfeel is neglected. Have you ever tried eating with your eyes closed? Try it, and think about what the food *feels* like in your mouth. You may be surprised to experience a variety of sensations that mimic hot and cold, as when eating something extremely spicy or something minty.

CHAPTER 40

Taste the Aroma

How do you taste what you smell?

IF YOU HAD TO LOSE ONE SENSE, which would it be? Sight? Hearing? Touch? Smell? Taste? The *Escapist* magazine once asked its readers this question in an online poll. Of the 539 people who responded, 10 would give up sight, 15 would give up hearing, 34 their sense of touch, 123 their sense of taste, and 357 their sense of smell.

Olfaction is the poor relation among our senses and is massively underestimated. Yet its marginal position in the realm of the senses is nothing new. Even Aristotle looked down upon the olfactory sense: 'Our sense of smell is weaker than that of any other creature, and it is the weakest of the human senses.' In contrast, he considered sight and hearing to be the 'philosophical senses' and as such the only ones useful for the obtaining of knowledge. Had the *Escapist* been able to ask Aristotle, he would probably not have hesitated to waive his sense of smell either.

We had previously assumed that the human nose could identify ten thousand different odors; today we know that it can probably distinguish more than a trillion different ones. By comparison, the human ear can recognise 340,000 audible tones, and our eyes can distinguish between 7.5 million different colours.

In an experiment conducted at the Rockefeller University in New York, researcher Andreas Keller and his team asked a group of volunteers to sniff out the faintest nuances differentiating similar odors. The participants were asked to identify the odd one out from among three mixtures. More than half of the test subjects were able to distinguish the different one when the mixtures overlapped by 75 per cent. Some were able to detect differences in mixtures even when they overlapped by 90 per cent. 'We have more sensitivity in our sense of smell than for which we give ourselves credit,' Keller says.

It is usually not until we lose the ability to smell—when we have a cold, for example—that we realise just how accurately our nose normally navigates us through our world. Not only do we feel insecure in social situations, but we are also constantly at risk of inhaling dangerous substances—think of smoke, mould, car exhaust, and household cleaners. The loss hits us particularly hard at mealtimes. We may barely be able to tell sweet from salty and sour from bitter or umami. But how would we be able to enjoy food without the aromas we absorb through the nose? No orange, vanilla, or cinnamon, and no freshly baked bread or fried bacon?

The French gastronome Brillat-Savarin noted: 'For myself, I am not only convinced that there is no full act of tasting without the participation of smell, but I am also tempted to believe that smell and taste form a single sense, of which the mouth is the laboratory and the nose is the chimney.'

Commonly when we talk about taste, we actually mean the smelling of aromas. But how exactly do we recognise aromas? When we exhale, scent molecules from solid or liquid foods travel with the air through the larynx (in our throat) into the nasal cavity (in our nose) where they hit the olfactory epithelium (skin-like tissue in our nose) and its numerous olfactory cells. There, each scent molecule latches onto a matching receptor cell like a key into a lock. So for the detection of smells it's the retronasal action—taking place at the back of our nose—of exhaling that's important rather than the anterior action of inhaling. If you want to enjoy maximum flavour intensity, make sure to chew your food carefully. The process of chewing releases additional scent molecules. Also remember that numerous foods are more flavoursome after they've been heated (see 'The Raw Food Fallacy', page 36). And finally, as culinary experts do, try slurping—it allows more air to get into the mouth to mix with the scent molecules.

Eventually the odour information passes from the olfactory epithelium to the various regions in the brain that are responsible for the control of emotions, memory, speech, and behaviour. In conjunction with information from our other senses of sight, hearing, and touch, it forms what we think of as the flavour of a dish, a process the leading US neuroscientist Gordon Shepherd calls the *human brain flavour system*. According to Shepherd, our specialised sense of smell plays a key part in it, as among all the factors that the brain uses to compile a flavour, smell is the most dominant.

No matter how scents find their way into our nasal cavity, be it through inhaling or exhaling, they usually play a subliminal role—we're not even aware of them until they cross a certain sensitivity threshold. Yet through the limbic system—the structure in our brain that controls our emotions—they still influence our behaviour. In South Korea, Dunkin' Donuts once installed scent modules in public transit buses that would spray a burst of coffee scent every time the company's jingle played on the PA system. Sales analyses revealed that in the shops located near the bus stations, sales spiked by 29 per cent. A casino in Las Vegas had already experimented with using fragrances in the nineties, with the result that 'odorised' slot machines increased player loyalty and hence the amount of money played at the machines compared to scent-neutral machines.

Smelling can also be learned and exercised. People with no or a very weak sense of smell, so-called anosmiacs, can learn to smell (better) by sniffing at special aromatic oils such as rose, clove, eucalyptus, and lemon twice a day. After only twelve weeks they should see significant improvement in their sense of smell. Over time, the number of olfactory receptors multiplies, the brain starts to restructure itself, and general odour perception improves. Regularly changing the odour range accelerates this regeneration. Since our ability to smell decreases with age, we can't start early enough to exercise our noses.

FOOD FOR THOUGHT: We may not live in an olfactory age, but our once-devalued sense of smell has found some new respect. We now know that it's intrinsically linked to our appreciation of a good meal. Contrary to popular belief, those without a sense of smell can still taste; however, they may have a diminished perception of flavour. If you were to reconsider the one sense you would lose, would anything be different?

CHAPTER 41

Chew on This

Why should you treat forks with caution?

WHAT DOES ELECTRICITY TASTE LIKE? Well, no one would voluntarily touch a live electric fence with their tongue; after all, we remember only too well the accidental shocks we received from them as children. Ultimately, however, the effective use of electricity is a matter of dosage. Hiromi Nakamura from the University of Tokyo has tested the boundaries with the development of an electric fork: 'We are inventing devices that add electricity to the tongue. We are trying to create virtual taste.'

Nakamura compares the way taste is created by electric stimulation of the tongue to the way sound is generated in the ear. Both are vibrations that can be produced artificially, and this is where we enter the terrain of so-called tech-cuisine, which examines the intersections between technology and food.

'Food hacking is about augmenting or diminishing real food,' Nakamura explains. 'It may seem like we're cooking, but we're actually working on the human senses.' It's not about technological gadgets for bored foodies, but about new solutions to nutritional problems such as high salt consumption. Nakamura has found that eating with an electric fork reduces the need for adding salt. That's because electricity not only tastes like fizzy champagne, it also intensifies, or, if you prefer, diminishes the perception of a salty taste.

The fork occupies a special place among our eating utensils. The knife and spoon have been part of our cooking and dining culture from the very beginning: Stone hand axes were used like knives to tear up meat, and seashells and nutshells were used like spoons to scoop liquids, differing from modern implements only in shape and material. The fork, however, is a comparatively new invention. Although the modern fork is among the most commonplace of objects, available in a range of shapes and forms such as the deli fork, fruit fork, salad fork, fish fork, ice cream fork, and dessert fork, its introduction was at first highly controversial. People preferred using their fingers to eat, a practice still common in many places around the world today. A fork should be used for 'taking food from the common dish', suggested Erasmus of Rotterdam around 1500 in his treatise on table manners. There was a real aversion to using a fork, which was considered 'effeminate' and 'a pointless affectation'. The first forks were small and used by Italian and French nobility for eating sweets and fruit. Henry III brought the fork to France in the sixteenth century and his courtiers were 'derided for this "affected" manner of eating', writes Norbert Elias in *The Civilizing Process*. Nor were they proficient at using a fork: 'It was said that half the food fell off the fork as it traveled from plate to mouth.' Not using one's fork to poke the food on one's plate, not inserting it too far into the mouth, and not waving it through the air when gesticulating are now understood as basic table manners, but not then. We may take them for granted, but they're something that 'had first to be slowly and laboriously acquired and developed by society as a whole'.

The introduction of a fork and table knife about 250 years ago changed not only our table manners but also the very formation of our mouths. The nineteenth-century anthropologist Charles Loring Brace was obsessed with finding an explanation for the development of the human overbite. In *Consider the Fork*, Bee Wilson writes, 'In premodern times, Brace surmises that the main method of eating would have been something he has christened "stuff-and-cut".' She describes the method as follows:

'First, grasp the food in one of your hands. Then clamp the end of it forcefully between your teeth. Finally, separate the main hunk of food from the piece in your mouth, either with a decisive tug of your hand or by using a cutting implement.' The real purpose of the incisors wasn't to cut food, Brace claimed, but to clamp it in the mouth. 'It is my suspicion,' suggested Brace, 'that if the incisors are used in such a manner several times a day . . . they will become positioned so that they normally occlude edge to edge.' With the introduction of the fork, however, we ceased needing to clamp our food between our teeth, and the top row stopped meeting the bottom.

Be that as it may, in tech-cuisine, the fork is very popular. Molecular gastronomists have developed something called the AROMAFORK, which you can purchase for about fifty dollars. It has a small circular recess that holds a piece of diffusing paper onto which a variety of aromatic flavors like basil, mint, truffle, banana, cinnamon, or coconut can be dropped. Can you trick your brain into thinking you're eating something that you're not? Apparently you *can*. Suddenly french fries taste of cinnamon and your steak has a strange hint of banana.

FOOD FOR THOUGHT: You've likely been taught that it's rude to eat with your fingers, but forks may be more trouble than they're worth. The list of fork faux pas is endless: don't hold it in the wrong hand or upside down; don't gesture with a fork in your hand; and, crucially, don't eat an oyster with a dinner fork. If, at some point in the future, the aroma fork becomes electric and 'champagne chicken' tastes of mint, it might be time to start eating with our fingers again.

CHAPTER 42

The Pineapple Fallacy

Why do you like what you like?

A PINEAPPLE IS A PINEAPPLE IS A PINEAPPLE? Right? No, not quite. The way you describe the precise taste of a pineapple, how sweet, juicy, stringy, soft, aromatic, and fizzy on the tongue you remember the fruit to be, depends on your personal taste profile. Your expectation of a pineapple's taste was shaped by when, and more importantly in which form (i.e., processed, packaged, or fresh), you first encountered the fruit. If your first pineapple came from a can and was buried between crust, sauce, ham, and cheese, you are a Hawaiian-pizza child. The advantage of this method of preparation is that it kills off the slightly metallic taste of the canned pineapple. People who have become conditioned to this artificial pineapple flavour, and who have come to love it, are in for a surprise when they taste fresh pineapple for the first time and wonder what's wrong with it.

Which foods we like and which ones we reject, which foods make regular appearances on our table and which ones never will, is determined by our socialisation within a certain food culture. If you grew up in a Thai village, you will naturally like different foods than someone whose childhood was spent eating cheese fondue and Bircher muesli in the Swiss mountains. And within each of these various foodscapes, personal likes and dislikes evolve, shaped by our education, learning processes, experiences, and genetic makeup (see 'Supertasters', page 178). We all go through the picky-eater phase. Neophobia, our innate fear of all things new, is particularly common in toddlers aged eighteen to twenty-four months. We remain largely sceptical until the age of five. This trait, which drives parents to distraction today, once ensured our survival.

The driving force behind our tastes is our family. Our parents, brothers, and sisters set a daily example of their taste preferences and thereby help shape our own—known as the *mere-exposure effect.* It's the phenomenon by which we tend to like things because we have become familiar with them. According to nutritional psychologist Volker Pudel, flavours that we have experienced tend to be repeated. This also offers a certain safety potential, he asserts, since the taste of a food helps us to identify something we have eaten and tolerated before without any negative consequences. When choosing a dish, we're then more likely to opt for one containing foods we've had before. And, one might add, because we have developed a liking for it, which is much more complex than simple repetition. The danger of repetition is that it can lead to sensory blunting. One of the reasons why our favourite meals are so special to us is that we don't eat them on a daily basis. This specific sensory saturation is based on an evolutionary mechanism that protects us from an unbalanced diet and thus from a lack of nutrients and vitamins. With one exception: breast milk.

However, even the scope of the mere-exposure effect has its limits, and sometimes it leads nowhere. If someone links a certain food to a negative experience, it is useless to offer him the offensive food again and again. Nonetheless, the mere-exposure effect can still be put to good use as an educational tool. Foods your offspring refuses to eat outright (and without reason) may well be tolerated once the child has been exposed to them often enough. The Welsh psychologist David Benton has put together some advice for parents on how to encourage healthy eating habits for their children:

- The emotional atmosphere of the meal is important. Do not use meal times as the opportunity to chastise, and do not let a child's failure to eat cause unpleasantness. . . .

- Forcing a child to eat a food will decrease the liking of that food. Neophobia is to be expected and should not be allowed to generate negativity.
- Encourage the child to be aware of satiety cues and allow these to define how much is eaten. If you wish a plate to be cleared, then allow the child to dictate the quantity placed on it or give a series of small servings until no more is needed.
- Parents should be careful that high-energy-density foods are not used as rewards and treats.

FOOD FOR THOUGHT: In the event that your attempts to follow this advice aren't wildly successful, it may help to remember how much you yourself had to learn, and are still learning, about the complexity of tastes. Just because you can't stand pineapple pizza, that doesn't mean it's repulsive to everyone.

Is there a food you haven't liked in the past, but think that you might? Try it in a different form or context! We're all a little bit picky at least some of the time. Some things take time; even taste experiences can be developed.

ACKNOWLEDGEMENTS

IT WAS A COOL AUTUMN EVENING as we dined in a wonderful Italian restaurant. Picture the scene: dimmed lights, soft music, glasses clinking, aromatic pasta, attentive waiters. A kind of *gesamtkunstwerk*, a piece of art addressing all the senses. It was nigh impossible not to enjoy one's food here. Everything was a promise. We tried to divide this promise into its individual parts. As we ate, we asked ourselves what exactly it was that made this dinner so coherent. Before long we were talking about psychology, and later on in the evening about our own culinary socialisation. Regarding our upbringing, we cover both ends of the healthy-eating spectrum: Melanie grew up on toasted sandwiches and TV dinners, while Diana grew up in the country where her family had a greenhouse and slaughtered their own animals. We were hooked on the subject, and the idea of this book was born. We researched countless studies and pored over scientific articles; we conducted interviews at the usual kitchen gatherings at parties. Our heartfelt thanks go to everyone who inspired us in their own way. Special thanks go to our editor, Batya Rosenblum, for her inspirational approach, to Carolin Sommer for her excellent translation, and to our families for their love and support. And last but not least we would like to thank all the chefs around the world who pursue their craft with passion and who conjure taste adventures that win a permanent place in our culinary memory.

BIBLIOGRAPHY

1. THE COLOUR OF FLAVOUR

Cannon, Dyan. *Dear Cary: My Life with Cary Grant*. New York: HarperCollins, 2011.

Duncker, Karl. 'The Influence of Past Experience upon Perceptual Properties'. *American Journal of Psychology* 52, no. 2 (April 1939): 255–65.

Harris, Gardiner. 'Colorless Food? We Blanch'. *The New York Times*. 2 April 2011. nytimes.com/2011/04/03/weekinreview/03harris.html.

Moir, H. C. 'Some Observations on the Appreciation of Flavour in Foodstuffs'. *Journal of Chemical Technology and Biotechnology* 55, no. 8 (February 1936): 145–48.

Petersen, C. *Referat zur Generalversammlung des Deutschen Milchwirtschaftlichen Vereins* [Lecture given at the general meeting of the Association of the German Dairy Industry]. Berlin, 1895.

Rozin, Paul. ' "Taste-Smell Confusions" and the Duality of the Olfactory Sense'. *Perception & Psychophysics* 31, no. 4 (July 1982): 397–401.

Shepherd, Gordon M. *Neurogastronomy: how the brain creates flavor and why it matters*. New York: Columbia University Press, 2011.

Spence, Charles. 'On the Psychological Impact of Food Colour'. *Flavour* 4, no. 21 (April 2015).

Zellner, Debra A., and Paula Durlach. 'Effect of Color on Expected and Experienced Refreshment, Intensity, and Liking of Beverages'. *American Journal of Psychology* 116, no. 4 (winter 2003): 633–47.

2. A PLATE OF ART

Freeman, Caitlin. *Modern Art Desserts: recipes for cakes, cookies, confections, and frozen treats based on iconic works of art*. Berkeley, CA: Ten Speed Press, 2013.

Harsdörffer, Georg Philipp. *Trincir-Buch*. Nürnberg: Fürst, 1652.

Michel, Charles, Carlos Velasco, Elia Gatti, and Charles Spence. 'A Taste of Kandinsky: assessing the influence of the artistic visual presentation of food on the dining experience'. *Flavour* 3, no. 7 (June 2014).

Redzepi, René. *Noma: time and place in nordic cuisine*. New York: Phaidon Press, 2010.

3. DISH DECISIONS

Genschow, Oliver, Leonie Reutner, and Michaela Wänke. 'The Color Red Reduces Snack Food and Soft Drink Intake'. *Appetite* 58, no. 2 (April 2012): 699–702.

Harrar, Vanessa, and Charles Spence. 'The Taste of Cutlery: how the taste of food is affected by the weight, size, shape, and colour of the cutlery used to eat it'. *Flavour* 2, no. 21 (June 2013).

Piqueras-Fiszman, Betina, Jorge Alcaide, Elena Roura, and Charles Spence. 'Is It the Plate or Is It the Food? Assessing the influence of the color (black or white) and shape of the plate on the perception of the food placed on it'. *Food Quality and Preference* 24, no. 1 (2012): 205–208.

Piqueras-Fiszman, Betina, Agnes Giboreau, and Charles Spence. 'Assessing the Influence of the Colour/Finish of the Plate on the Perception of the Food in a Test in a Restaurant Setting'. *Flavour* 2, no. 24 (August 2013).

Spence, Charles, and Betina Piqueras-Fiszman. *The Perfect Meal: the multisensory science of food and dining*. New York: John Wiley & Sons, 2014.

Van Ittersum, Koert, and Brian Wansink. 'Plate Size and Color Suggestibility: the Delboeuf illusion's bias on serving and eating behavior'. *Journal of Consumer Research* 39, no. 2 (August 2012): 215–28.

4. ALL YOU CAN ~~EAT~~ SEE

Foo, H., and Peggy Mason. 'Analgesia Accompanying Food Consumption Requires Ingestion of Hedonic Foods'. *Journal of Neuroscience* 29, no. 41 (October 2009): 13053–62.

Stein-Hölkeskamp, Elke. *Das römische Gastmahl: eine kulturgeschichte* (*The Roman Feast: a cultural history*). Munich: C. H. Beck, 2005.

Wansink, Brian. *Mindless Eating: why we eat more than we think*. New York: Bantam, 2006.

Wansink, Brian, and Collin R. Payne. 'Eating Behavior and Obesity at Chinese Buffets'. *Obesity* 16, no. 8 (August 2008).

5. SUPERMARKET SCHEMES

Dobelli, Rolf. *The Art of Thinking Clearly*. Translated by Nicky Griffin. New York: HarperCollins, 2013.

Lindstrom, Martin. *Brandwashed: tricks companies use to manipulate our minds and persuade us to buy*. New York: Crown Publishing, 2011.

Packard, Vance. *The Hidden Persuaders*. London: Longmans, 1962.

Pfaff, Jan. 'Das Unterbewusstsein kauft ein' ('The Subconscious Mind Buys'). *Der Freitag* (Berlin). September 30, 2009. freitag.de/autoren/jan-pfaff/das-unterbewusstsein-kauft-ein.

6. CELEBRITY-ADVICE FAIRY TALES

Cederström, Carl, and André Spicer. *The Wellness Syndrome.* Cambridge: Polity Press, 2015.

Kitz, Volker, and Manuel Tusch. *Psycho? Logisch! Nützliche Erkenntnisse aus der Alltagspsychologie* (*Psycho? Logical! Useful Insights from Everyday Psychology*). Munich: Wilhelm Heyne Verlag, 2011.

7. THE RAW-FOOD FALLACY

Kessler, David A. *The End of Overeating: taking control of the insatiable American appetite.* New York: Rodale Books, 2009.

Lévi-Strauss, Claude. *The Raw and the Cooked.* Vol. 1 of *Mythologiques.* Translated by John and Doreen Weightman. Chicago: University of Chicago Press, 1983.

Moss, Michael. *Salt Sugar Fat: how the food giants hooked us.* New York: Random House, 2014.

Pellegrini, Nicoletta, Emma Chiavaro, Claudio Gardana, Teresa Mazzeo, Daniele Contino, Monica Gallo, Patrizia Riso, Vincenzo Fogliano, and Marisa Porrini. 'Effect of Different Cooking Methods on Color, Phytochemical Concentration, and Antioxidant Capacity of Raw and Frozen *Brassica* Vegetables'. *Journal of Agricultural and Food Chemistry* 58, no. 7 (April 2010): 4310–21.

Rosati, Alexandra G., and Felix Warneken. 'Cognitive Capacities for Cooking in Chimpanzees'. *Proceedings of the Royal Society B* 282, no. 1809 (June 2015).

Shepherd. *Neurogastronomy.*

Washburn, S. L., ed. *Social Life of Early Man.* Oxfordshire: Routledge, 2004. First published 1962 by Wenner-Gren Foundation for Anthropological Research.

———, ed. *Classification and Human Evolution*. Oxfordshire: Routledge, 2004. First published 1964 by Wenner-Gren Foundation for Anthropological Research.

———. 'Evolution of Human Behavior'. In *Behavior and Evolution*, edited by Anne Roe and George Gaylord Simpson. New Haven: Yale University Press, 1958.

Wrangham, Richard. *Catching Fire: how cooking made us human*. New York: Basic Books, 2009.

8. GLUTEN ANXIETY

Davis, William. *Wheat Belly: lose the wheat, lose the weight, and find your path back to health*. New York: Rodale, 2011.

Laass, M. W., et al. 'Zöliakieprävalenz bei Kindern und Jugendlichen in Deutschland [Prevalence of celiac disease in children and teenagers in Germany]'. *Deutsches Ärzteblatt*, Vol. 12 (33–34) (2015): 553–560.

'Pedestrian Question—What Is Gluten?' YouTube video. 3:49. From *Jimmy Kimmel Live*. Posted by 'Jimmy Kimmel Live', 6 May 2014. youtube.com/watch?v=AdJFE1sp4Fw.

Pollmer, Udo. 'Bestrahlte Lebensmittel: wie unser Essen mit Atomtechnik in Berührung kommt' ('Irradiated Groceries: how our food comes into contact with nuclear technology'). *Deutschlandfunk Kultur*, 20 March 2011, deutschlandfunkkultur.de/bestrahlte-lebensmittel.993.de.html?dram:article_id=154559.

Schoenfeld, Jonathan D., and John P. A. Ioannidis. 'Is Everything We Eat Associated with Cancer? A systematic cookbook review'. *American Journal of Clinical Nutrition* 97, no. 1 (January 2013): 127–34.

Ventura, Alessandro, Giuseppe Magazzù, and Luigi Greco. 'Duration of Exposure to Gluten and Risk for Autoimmune Disorders in Patients with Celiac Disease'. For the SIGEP Study Group for Autoimmune Disorders in Celiac Disease. *Gastroenterology* 117, no. 2 (August 1999): 297–303.

9. CARB PHOBIA

Pollan, Michael. *In Defense of Food: an eater's manifesto*. New York: Penguin Press, 2008.

10. 'LOSE TEN POUNDS FAST!'

Cell Press. ' "Healthy" Foods Differ by Individual'. EurekAlert! 19 November 2015. eurekalert.org/pub_releases/2015-11/cp-fd111215.php.

Duhigg, Charles. *The Power of Habit: why we do what we do in life and business*. New York: Random House, 2012.

Fedoroff, Ingrid, Janet Polivy, and C. Peter Herman. 'The Specificity of Restrained Versus Unrestrained Eaters' Responses to Food Cues: general desire to eat, or craving for the cued food?' *Appetite* 41, no. 1 (August 2003): 7–13.

Mann, Traci. *Secrets from the Eating Lab: the science of weight loss, the myth of willpower, and why you should never diet again*. New York: Harper Wave, 2015.

Rein, Michal, Gili Zilberman-Schapira, Lenka Dohnalová, Meirav Pevsner-Fischer, Rony Bikovsky, Zamir Halpern, Eran Segal, et al. 'Personalized Nutrition by Prediction of Glycemic Responses'. *Cell* 163, no. 5 (November 2015): 1079–94.

11. ENOUGH IS ENOUGH

Brunstrom, Jeffrey M., Jane Collingwood, and Peter J. Rogers. 'Perceived Volume, Expected Satiation, and the Energy

Content of Self-Selected Meals'. *Appetite* 55, no. 1 (August 2010): 25–29.

Brunstrom, Jeffrey M., and Peter J. Rogers. 'How Many Calories Are on Our Plate? Expected fullness, not liking, determines meal-size selection'. *Obesity* 17, no. 10 (October 2009): 1884–90.

Colagiuri, Ben, and Peter F. Lovibond. 'How Food Cues Can Enhance and Inhibit Motivation to Obtain and Consume Food'. *Appetite* 84, no. 1 (January 2015): 79–87.

Dailey, Megan J., Timothy H. Moran, Peter C. Holland, and Alexander W. Johnson. 'The Antagonism of Ghrelin Alters the Appetitive Response to Learned Cues Associated with Food'. *Behavioural Brain Research* 303 (April 2016): 191–200.

Enders, Giulia. *Gut: the inside story of our body's most underrated organ*. Translated by David Shaw. Vancouver: Greystone Books, 2015.

Geliebter, A. 'Gastric Distension and Gastric Capacity in Relation to Food Intake in Humans'. *Physiology & Behavior* 44, no. 4–5 (1988): 665–68.

Geliebter, A., S. Westreich, and D. Gage. 'Gastric Distention by Balloon and Test-Meal Intake in Obese and Lean Subjects'. *American Journal of Clinical Nutrition* 48, no. 3 (September 1988): 592–94.

Nguo, K., K. Z. Walker, M. P. Bonham and C. E. Huggins. 'Systematic Review and Meta-Analysis of the Effect of Meal Intake on Postprandial Appetite-Related Gastrointestinal Hormones in Obese Children'. *International Journal of Obesity* 40, no. 4 (April 2016): 555–63.

Oldham-Cooper, Rose E., Charlotte A. Hardman, Charlotte E. Nicoll, Peter J. Rogers, and Jeffrey M. Brunstrom. 'Playing a Computer Game During Lunch Affects Fullness, Memory for Lunch, and Later Snack Intake'. *American Journal of Clinical Nutrition* 93, no. 2 (February 2011): 308–13.

Rozin, Paul, Sara Dow, Morris Moscovitch, and Suparna Rajaram. 'What Causes Humans to Begin and End a Meal? A role for memory for what has been eaten, as evidenced by a study of multiple meal eating in amnesic patients'. *Psychological Science* 9, no. 5 (1998): 392–96.

Stevenson, Richard J., and John Prescott. 'Human Diet and Cognition'. *Wiley Interdisciplinary Reviews Cognitive Science* 5, no. 4 (July/August 2014): 463–75.

Wansink, Brian, and Matthew M. Cheney. 'Super Bowls: serving bowl size and food consumption'. *Journal of the American Medical Association* 293, no. 14 (April 2005): 1727–28.

Woods, Stephen C. 'Gastrointestinal Satiety Signals I. An overview of gastrointestinal signals that influence food intake'. *American Journal of Physiology—Gastrointestinal and Liver Physiology* 286, no. 1 (January 2004): G7–G13.

12. LOVE, THE ANTI-DIET

Brillat-Savarin, Jean Anthelme. *The Physiology of Taste: or meditations on transcendental gastronomy*. Translated by MFK Fisher. New York: Vintage Press, 2009.

Kaufmann, Jean-Claude. *The Meaning of Cooking*. Cambridge: Polity Press, 2010.

Kellow, Juliette. 'Is Your Partner Making You Fat?' *Weight Loss Resources*. Accessed 16 May 2017. weightlossresources .co.uk/healthy_eating/living_together.htm.

Leimgruber, Walter. 'Zwischen Fasten und Völlerei: Essen und Trinken als Thema der Kulturwissenschaft' ('Between Fasting and Gluttony: eating and drinking in cultural science'). Basel: Science Lunch lecture, 2005.

'The War of the Roses—Trailer (1989)'. YouTube video. 2:35. *The War of the Roses* theatrical trailer. Posted by 'WorleyClarence,' 15 June 2008. youtube.com/watch?v=5ebv3i_9Ltc.

13. THE DOGGIE-BAG PARADOX

Amer, Stacey, and Caroline McClatchey. 'Doggy Bag: why are the British too embarrassed to ask?' *BBC News Magazine*. 5 October 2011. bbc.com/news/magazine-15106212.

Breeden, Aurelien. 'Brushing Off a French Stigma That Doggie Bags Are for Beggars'. *The New York Times*. 13 November 2014. nytimes.com/2014/11/14/world/europe/brushing-off-a-french-stigma-that-doggie-bags-are-for-beggars-.html?_r=0.

Davidson, Alan. *The Oxford Companion to Food*. Edited by Tom Jaine. Oxford: Oxford University Press, 2014.

Spencer, Colin. *British Food: an extraordinary thousand years of history*. London: Grub Street Cookery, 2011.

14. MUSIC TO YOUR . . . STOMACH

Carvalho, Felipe Reinoso, Raymond Van Ee, Monika Rychtarikova, Abdellah Touhafi, Kris Steenhaut, Dominique Persoone, and Charles Spence. 'Using Sound-Taste Correspondences to Enhance the Subjective Value of Tasting Experiences'. *Frontiers in Psychology* 6 (2015): 1309.

Crisinel, Anne-Sylvie, and Charles Spence. 'Implicit Association Between Basic Tastes and Pitch'. *Neuroscience Letters* 464, no. 1 (October 2009): 38–42.

'Egg and Bacon Ice Cream'. YouTube video. 1:56. Heston Blumenthal describes how to make egg and bacon ice cream on *Kitchen Chemistry*. Posted by 'rikaitch', 13 April 2016. youtube.com/watch?v=D6CLoRuvGcY.

Eplett, Layla. 'Pitch/Fork: the relationship between sound and taste'. *Scientific American*, 4 September 2013. blogs.scientificamerican.com/food-matters/pitchfork-the-relationship-between-sound-and-taste.

Simner, Julia, Christine Cuskley, and Simon Kirby. 'What Sound Does That Taste? Cross-modal mappings across gustation and audition'. *Perception* 39, no. 4 (2010): 553–69.

Spence and Piqueras-Fiszman. *The Perfect Meal.*

Spence, Charles, and Maya U. Shankar. 'The Influence of Auditory Cues on the Perception of, and Responses to, Food and Drink'. *Journal of Sensory Studies* 25, no. 3 (June 2010): 406–30.

Twilley, Nicola. 'Accounting for Taste'. *The New Yorker.* 2 November 2015. newyorker.com/magazine/2015/11/02/accounting-for-taste.

Wang, Qian (Janice), Sheila Wang, and Charles Spence. '"Turn Up the Taste": assessing the role of taste intensity and emotion in mediating crossmodal correspondences between basic tastes and pitch'. *Chemical Senses* 41, no. 4 (February 2016): 345–56.

Wheeler, Elmer. *Tested Sentences That Sell: how to use 'word magic' to sell more and work less!*. New York: Prentice Hall Inc., 1937.

Woods, A. T., E. Poliakoff, D. M. Lloyd, J. Kuenzel, R. Hodson, H. Gonda, J. Batchelor, G. B. Dijksterhuis, and A. Thomas. 'Effect of Background Noise on Food Perception'. *Food Quality and Preference* 22, no. 1 (January 2011): 42–47.

15. SMACK-AND-SLURP PHOBIA

Bernstein, Elizabeth. 'Annoyed by Loud Chewing? The Problem Is You'. *The Wall Street Journal.* 19 October 2015. wsj.com/articles/annoyed-by-loud-chewing-the-problem-is-you-1445277757.

Dozier, Thomas H. *Understanding and Overcoming Misophonia: a conditioned aversive reflex disorder.* Livermore, CA: Misophonia Treatment Institute, 2015.

Fitzmaurice, Guy. 'The Misophonia Activation Scale'. *Misophonia UK*. Accessed 16 May 2017. misophonia-uk.org/the-misophonia-activation-scale.html.

Kopp, Diana von. 'Oh du schreckliches Weihnachtsessen' ('Tidings of Discomfort and Noise'). *Food Affair* (blog). blogs.faz.net/foodaffair/2015/12/23/oh-du-schreckliches-weihnachtsessen-628.

Schröder, Arjan, Nienke Vulik, and Damiaan Denys. 'Misophonia: diagnostic criteria for a new psychiatric disorder'. *PLOS ONE* 8, no. 1 (January 2013): e54706.

Webber, Troy A., and Eric A. Storch. 'Toward a Theoretical Model of Misophonia'. *General Hospital Psychiatry* 37, no. 4 (July–August 2015): 369–70.

Wu, Monica S., Adam B. Lewin, Tanya K. Murphy, and Eric A. Storch. 'Misophonia: incidence, phenomenology, and clinical correlates in an undergraduate student sample'. *Journal of Clinical Psychology* 70, no. 10 (October 2014): 994–1007.

16. THE FLAVOUR OF MUSIC

Crisinel, Anne-Sylvie, Stefan Cosser, Scott King, Russ Jones, James Petrie, and Charles Spence. 'A Bittersweet Symphony: systematically modulating the taste of food by changing the sonic properties of the soundtrack playing in the background'. *Food Quality and Preference* 24, no. 1 (April 2012): 201–204.

Eplett, Layla. 'The Sound (and Taste) of Music'. *Scientific American* (blog). 9 December 2014. blogs.scientificamerican.com/food-matters/the-sound-and-taste-of-music.

Fink, Hans-Juergen. 'Elmar Lampson hat schon als Kind in Tönen geträumt' ('Elmar Lampson Dreamt in Sounds Even as a Child'). *Hamburger Abendblatt*. 25 September 2013. abendblatt.de/kultur-live/article120359547/Elmar-Lampson-hat-schon-als-Kind-in-Toenen-getraeumt.html.

FOCUS Online. 'Premiere in Hamburg: Töne beeinflussen Weingeschmack' ('Premiere in Hamburg: sounds affect the taste of wine'). 9 March 2015. focus.de/reisen/reise-news/wein-premiere-in-hamburg-toene-beeinflussen-weingeschmack_id_4531248.html.

North, Adrian C., and David J. Hargreaves. 'The Effects of Music on Responses to a Dining Area'. *Journal of Environmental Psychology* 16, no. 1 (March 1996): 55–64.

Steiner, Paul. *Sound Branding, Grundlagen der akustischen Markenführung* (*Sound Branding: principles of acoustic brand leadership*). Wiesbaden: Gabler, 2009.

Strobele, Nanette, and John M. de Castro. 'Listening to Music While Eating Is Related to Increases in People's Food Intake and Meal Duration'. *Appetite* 47, no. 3 (November 2006): 285–89.

Victor, Anucyia. 'Louis Armstrong for Starters, Debussy with Roast Chicken and James Blunt for Dessert: British Airways pairs music to meals to make in-flight food taste better'. *Daily Mail*. 15 October 2014, dailymail.co.uk/travel/travel_news/article-2792286/british-airways-pairs-music-meals-make-flight-food-taste-better.html.

17. THE PERFECT CRISP

Beckerman, Joel. *The Sonic Boom: how sound transforms the way we think, feel, and buy*. New York: Houghton Mifflin Harcourt, 2014.

Bennett, Lynne. 'Fun Facts About Frites'. *San Francisco Chronicle*. 20 September 2000. sfgate.com/recipes/article/Fun-Facts-About-Frites-2738230.php.

Bethge, Philip, Jörg Blech, Rafaela von Bredow, Nils Klawitter, Julia Koch, Udo Ludwig, Christoph Schult, and Samiha Shafy. 'Die Menschen-Mäster' ('The People Fatteners'). *Der Spiegel* 10. 4 March 2013. spiegel.de/spiegel/print/d-91346600.html.

Burhans, Dirk. *Crunch! A history of the great American potato chip*. Madison, WI: Terrace Books, 2008.

Kitchiner, William. *The Cook's Oracle*. 4th ed. Edinburgh, London: A. Constable and Co., 1822.

Original Saratoga Chips. 'The Story of George Crum and America's First Kettle Chip'. Accessed on 16 May 2017. originalsaratogachips.com/our-story.

Spence, Charles, Maya U. Shankar, and Heston Blumenthal. ' "Sound Bites": auditory contributions to the perception and consumption of food and drink'. In *Art and the Senses*, edited by Francesca Bacci and David Melcher, 207–38. Oxford: Oxford University Press, 2011.

Vranica, Suzanne. 'Snack Attack: chip eaters make noise about a crunchy bag'. *The Wall Street Journal*. 18 August 2010. wsj.com/articles/SB10001424052748703960004575427150103293906.

Zampini, Massimiliano, and Charles Spence. 'The Role of Auditory Cues in Modulating the Perceived Crispness and Staleness of Potato Chips'. *Journal of Sensory Studies* 19, no. 5 (October 2004): 347–63.

18. THE 'UNHEALTHY = TASTY INTUITION'

Mai, Robert, and Stefan Hoffman. 'How to Combat the Unhealthy = Tasty Intuition: the influencing role of health consciousness'. *Journal of Public Policy & Marketing*, 34, no. 1 (spring 2015): 63–83.

Raghunathan, Rajagopal, Rebecca Walker Naylor, and Wayne D. Hoyer. 'The Unhealthy = Tasty Intuition and Its Effects on Taste Inferences, Enjoyment, and Choice of Food Products'. *Journal of Marketing* 70, no. 4 (October 2006): 170–84.

Sütterlin, Bernadette, and Michael Siegrist. 'Simply Adding the Word "Fruit" Makes Sugar Healthier: the misleading effect of symbolic information on the perceived healthiness of food'. *Appetite* 95, no. 1 (December 2015): 252–61.

Werle, Carolina O. C., Olivier Trendel, and Gauthier Ardito. 'Unhealthy Food Is Not Tastier for Everybody: the 'healthy = tasty' French intuition'. *Food Quality and Preference* 28, no. 1 (April 2013): 116–21.

Witherly, Steven A. *Why Humans Like Junk Food: the inside story on why you like your favorite foods, the cuisine secrets of top chefs, and how to improve your own cooking without a recipe!*. Lincoln, NE: iUniverse, 2007.

19. STATUS ANXIETY À LA CARTE

Aiken, Kristen. 'Chefs Confess That "If You Don't Tell People What They're Eating, They'll Like It" '. *Huffington Post*. 9 April 2014. huffingtonpost.com/2014/04/09/chef-menu-secrets_n_5120363.html.

Ariely, Dan. *Predictably Irrational: the hidden forces that shape our decisions*. New York: HarperCollins, 2009.

Jurafsky, Dan. *The Language of Food: a linguist reads the menu*. New York: W. W. Norton & Company, 2014.

20. 'HOW CAN YOU EAT *THAT?*'

Curtis, Val, Robert Aunger, and Tamer Rabie. 'Evidence That Disgust Evolved to Protect from Risk of Disease'. *Procedings of the Royal Society B: Biological Sciences* 271, sup. 4 (May 2004): S131–S133.

Curtis, Valerie. 'Why Disgust Matters'. *Philosophical Transactions of the Royal Society B: Biological Sciences* 366, no. 1583 (December 2011): 3478–90.

Harris, Marvin. *Good to Eat: riddles of food and culture.* Long Grove, IL: Waveland Press, 1998.

Mennell, Stephen. *All Manners of Food: eating and taste in England and France from the middle ages to the present.* Champaign, IL: Illini Books, 1996.

Reitmeier, Simon. *Warum wir mögen, was wir essen: Eine Studie zur Sozialisation der Ernährung* (*Why We Like What We Eat: a study of the socialisation of food*). Bielefeld: transcript Verlag, 2013.

Rozin, Paul, Jonathan Haidt, and Clark R. McCauley. 'Disgust'. In *The Oxford Companion to Emotion and the Affective Sciences*, edited by David Sander and Klaus R. Scherer, 121–122. Oxford: Oxford University Press, 2009.

Tücke, Manfred. *Grundlagen der Psychologie für (zukünftige) Lehrer* (*Principles of Psychology for Teachers [to Be]*). Berlin: LIT Verlag, 2003.

Von Engelhardt, Dietrich. *Geschmackskulturen: Vom Dialog der Sinne beim Essen und Trinken* (*Taste Cultures: how our senses communicate when we eat and drink*). Edited by Rainer Wild. Frankfurt: Campus Verlag, 2005.

21. THE MARKETING-PLACEBO EFFECT

Hay, Colin. 'When Points Mean Prices'. *Decanter.* 21 August 2008. decanter.com/features/when-points-mean-prices-247008.

Kirby, Terry. 'Robert Parker Interview: the world's top wine critic on tasting 10,000 bottles a year, absurd drinking notes and new world wannabes'. *Independent.* 29 March 2015. independent.co.uk/life-style/food-and-drink/robert-parker-interview-the-worlds-top-wine-critic-on-tasting-10000-bottles-a-year-absurd-drinking-10135796.html.

Parker, Robert M. Jr. *Parker's Wine Buyer's Guide: the complete, easy-to-use reference on recent vintages, prices, and ratings for more than 8,000 wines from all the major wine regions.* 7th ed. New York: Simon & Schuster, 2008.

Plassmann, Hilke, and Bernd Weber. 'Individual Differences in Marketing Placebo Effects: evidence from brain imaging and behavioral experiments'. *Journal of Marketing Research* 52, no. 4 (August 2015): 493–510.

Riedl, Joachim. 'Der große Grand-Cru-Schwindel' ('The Great Grand-Cru Fraud'). *Falstaff.* 24 November 2013. falstaff.at/nd/der-grosse-grand-cru-schwindel.

Solomon, Gregg Eric Arn. 'Psychology of Novice and Expert Wine Talk'. *American Journal of Psychology* 103, no. 4 (winter 1990): 495–517.

22. THE PRIMING EFFECT

Aarts, Henk, Ap Dijksterhuis, and Peter de Vries. 'On the Psychology of Drinking: being thirsty and perceptually ready'. *British Journal of Psychology* 92, no. 4 (November 2001): 631–42.

Bargh, John A., Mark Chen, and Lara Burrows. 'Automaticity of Social Behavior: direct effects of trait construct and stereotype activation on action'. *Journal of Personality and Social Psychology* 71, no. 2 (August 1996): 230–44.

Brouwer, Amanda, and Katie Mosack. 'Motivating Healthy Diet Behaviours: the self-as-doer identity'. *Self and Identity* 14, no. 6 (May 2015): 638–53.

Kahneman, Daniel. *Thinking, Fast and Slow.* New York: Farrar, Straus and Giroux, 2011.

Karremans, Johan C., Wolfgang Stroebe, and Jasper Claus. 'Beyond Vicary's Fantasies: the impact of subliminal priming and brand choice'. *Journal of Experimental Social Psychology* 42, no. 6 (November 2006): 792–98.

Key, Wilson Bryan. *Subliminal Seduction*, Englewood Cliffs, NJ: Signet, 1973.

Packard. *The Hidden Persuaders.*

Traufetter, Gerald. 'Stimme aus dem Nichts' ('Voice out of Nowhere'). *Der Spiegel* 15. 10 April 2006. spiegel.de/spiegel/print/d-46581582.html.

Wansink, Brian. *Slim by Design: mindless eating solutions for everyday life.* New York: HarperCollins, 2014.

23. THE HEALTH-HALO EFFECT

Baumgartner, Hans, and Joerg Koenigstorfer. 'The Effect of Fitness Branding on Restrained Eaters' Food Consumption and Postconsumption Physical Activity'. *Journal of Marketing Research* 53, no. 1 (February 2016): 124–38.

Comish, Chris. 'UH Researchers Warn: food marketing wizards using 'health trigger' words to mislead consumers'. *BioNews Texas.* 16 June 2004. bionews-tx.com/news/2014/06/16/uh-researchers-warn-food-marketing-wizards-using-health-trigger-words-to-mislead-consumers.

Demircan, Ozan. 'Verordneter Strukturwandel' ('Structural Change by Decree'). *Handelsblatt* 25 (2016): 32.

Linder, N. S., Gabriele Uhl, Klaus Fliessbach, P. Trautner, Christian E. Elger, and Bernd Weber. 'Organic Labeling Influences Food Valuation and Choice'. *Neuroimage* 53, no. 1 (October, 2010): 215–20.

Northup, Temple. 'Truth, Lies, and Packaging: how food marketing creates a false sense of health'. *Food Studies: an interdisciplinary journal* 3, no. 1 (March 2014): 9–18.

Pollan. *In Defense of Food.*

———. *Food Rules: an eater's manual.* New York: Penguin Press, 2009.

Technische Universität München. 'Prof. Königstorfer-Studie: "Fitness"-Label als gefährlicher Heiligenschein' ('Prof. Königstorfer study: "Fitness" label as dangerous halo'). Fakultät für Sport- und Gesundheitswissenschaften. 29 July 2015. www.sg.tum.de/news/news-singleview-fakultaet/article/prof-koenigstorfer-studie-fitness-label-als-gefaehrlicher-heiligenschein.

24. THE ROMEO-AND-JULIET EFFECT

Brehm, Sharon S., and Jack W. Brehm. *Psychological Reactance: a theory of freedom and control.* New York: Academic Press, 1981.

Burger, Jörg. 'Eltern, hört endlich auf, von gesundem Essen zu reden! Wie man Kämpfe am Esstisch vermeidet: Ein Gespräch mit dem Ernährungspsychologen Thomas Ellrott' ('Parents, Stop Talking About Healthy Eating! How to Avoid Battles at the Dinner Table: an interview with nutritional psychologist Thomas Ellrott'). *Zeit* 17. 20 April 2011. zeit.de/2011/17/Genuss-Interview.

Carpenter, Christopher J. 'A Meta-Analysis of the Effectiveness of the "But You Are Free" Compliance-Gaining Technique'. *Communication Studies* 64, no. 1 (January 2013): 6–17.

Eyal, Nir. 'Why Behavior Change Apps Fail to Change Behavior'. *Nir & Far.* July 2013. nirandfar.com/2013/07/why-behavior-change-apps-fail-to-change-behavior.html.

Finkelstein, Stacey R., and Ayelet Fishbach. 'When Healthy Food Makes You Hungry'. *Journal of Consumer Research* 37, no. 3 (October 2010): 357–67.

Grynbaum, Michael M. 'New York's Ban on Big Sodas Is Rejected by Final Court'. *The New York Times.* 26 June 2014. nytimes.com/2014/06/27/nyregion/city-loses-final-appeal-on-limiting-sales-of-large-sodas.html.

Hanks, Andrew S., David R. Just, and Brian Wansink. 'Chocolate Milk Consequences: a pilot study evaluating the consequences of banning chocolate milk in school cafeterias'. *PLOS ONE* 9, no. 4 (April 2014): e91022.

Healy, Melissa. 'Proposed Soda Ban Likely to Backfire, Study Finds'. *Los Angeles Times.* 11 April 2013. articles.latimes.com/2013/apr/11/science/la-sci-small-sodas-20130411.

Schell, Jesse. *The Art of Game Design: a book of lenses.* Burlington, MA: Morgan Kaufmann Publishers, 2008.

Susher, Jacob, Raj Raghunathan, and Wayne Hoyer. 'Eating Healthy or Feeling Empty? how the 'healthy = less filling' intuition influences satiety'. *Journal of the Association for Consumer Research* 1, no. 1 (January 2016): 26–40.

Wansink, Brian, and David Just. 'Soda Ban Will Fail and Jeopardize Future Public Health Efforts'. Debate Club. *US News and World Report.* 1 June 2012. usnews.com/debate-club/should-the-sale-of-large-sugary-drinks-be-prohibited/soda-ban-will-fail-and-jeopardize-future-public-health-efforts.

Wilson, Brent M., Stephanie Stolarz-Fantino, and Edmund Fantino. 'Regulating the Way to Obesity: unintended consequences of limiting sugary drink sizes'. *PLOS ONE* 8, no. 4 (April 2013): e61081.

25. SLEEP YOURSELF SLIM

Brillat-Savarin. *The Physiology of Taste.*

Koch, Susanne. '5 Fragen an Prof. Dr. Jürgen Zulley' ('5 Questions for Dr. Jürgen Zulley'). *Report Psychologie.* 5 October 2011. report-psychologie.de/fileadmin/thema/2011/10/5_Fragen_an_Prof_Dr_Juergen_Zulley.pdf.

Psychologie Heute. *Futter für die Seele: Wie Gefühle uns beim Essen steuern—und warum Genuss ohne Reue möglich ist (Food for the Soul: how our feelings control how and what we eat—and why it is possible to enjoy food without regret).* Compact 44. Weinheim: Beltz Verlag, 2016.

St-Onge, Marie-Pierre, Amy L. Roberts, Jinya Chen, Michael Kelleman, Majella O'Keeffe, Arindam RoyChoudhury, and Peter J. H. Jones. 'Short Sleep Duration Increases Energy Intakes but Does Not Change Energy Expenditure in Normal-Weight Individuals'. *American Journal of Clinical Nutrition* 94, no. 2 (August 2011): 410–16.

St-Onge, Marie-Pierre, Amy L. Roberts, Ari Shechter, and Arindam RoyChoudhury. 'Fiber and Saturated Fat Are Associated with Sleep Arousals and Slow Wave Sleep'. *Journal of Clinical Sleep Medicine* 12, no. 1 (January 2016): 19–24.

26. THE FEEDING CLOCK

Cell Press. 'What You Eat May Affect Your Body's Internal Biological Clock'. ScienceDaily. 10 July 2014. sciencedaily.com/releases/2014/07/140710130852.htm.

Ehret, Charles F., and Lynne Waller Scanlon. *Overcoming Jet Lag.* New York: Berkley Books, 1983.

Fraunhofer-Institut für Bauphysik. 'Rätsel um Tomatensaft im Flugzeug gelöst' ('Puzzle of Tomato Juice on Airplanes Solved'). Accessed 16 May 2017. ibp.fraunhofer.de/de/Presse_und_Medien/Presseinformationen/Raetsel_um_Tomatensaftgeloest.html.

Sato, Miho, Mariko Murakami, Koichi Node, Ritsuko Matsumura, and Makoto Akashi. 'The Role of the Endocrine System in Feeding-Induced Tissue-Specific Circadian Entrainment'. *Cell Reports* 8, no. 2 (July 2014): 393–401.

Spence, Charles, Charles Michel, and Barry Smith. 'Airplane Noise and the Taste of Umami'. *Flavour* 3, no. 2 (February 2014).

27. 'I'LL HAVE WHAT YOU'RE HAVING'

Ariely. *Predictably Irrational.*

Ariely, Dan, and Jonathan Levav. 'Sequential Choice in Group Settings: taking the road less traveled and less enjoyed'. *Journal of Consumer Research* 27, no. 3 (December 2000): 279–90.

Döring, Tim, and Brian Wansink. 'The Waiter's Weight: does a server's BMI relate to how much food diners order?' *Environment and Behavior* 49, no. 2 (February 2017): 192–214.

Ellison, Brenna, and Jayson Lusk. ' "I'll Have What He's Having": group ordering behavior in food choice decisions'. *Food Quality and Preference* 37 (October 2014): 79–86.

28. NUDGING

Kroese, Floor M., Marchiori, David, and Denise de Ridder. 'Nudging Healthy Food Choices: a field experiment at the train station'. *Journal of Public Health Advance Access* 38, no. 2 (June 2015): 1–5.

Mühl, Melanie. 'Wie Google das Gewicht seiner Mitarbeiter kontrolliert' ('How Google Controls the Weight of Their Staff'). *Food Affair* (blog). *Frankfurter Allgemeine Zeitung.* 3 June 2015. blogs.faz.net/foodaffair/2015/06/03/wie-google-das-gewicht-seiner-mitarbeiter-kontrolliert-157.

Steel, Tanya. 'Inside Google's Kitchens'. *Gourmet*. 7 March 2012. gourmet.com/food/gourmetlive/2012/030712/inside-googles-kitchens.html.

Thaler, Richard H., and Cass R. Sunstein. *Nudge: improving decisions about health, wealth, and happiness*. New York: Penguin Books, 2009.

Wansink. *Slim by Design*.

29. THE FOOD RADIUS

Wansink. *Slim by Design*.

Winterhalder, Bruce, and Eric Alden Smith, eds. *Hunter-Gatherer Foraging Strategies: ethnographic and archeological analyses*. Chicago: University of Chicago Press, 1981.

30. THE TROPHY-KITCHEN SYNDROME

Albala, Ken, ed. *The SAGE Encyclopedia of Food Issues*. New York: SAGE Publications, 2015.

Bernstein, Fred A. 'The New Kitchen Is Done. So Why Can't I Be Happy?' *The New York Times*. 22 February 2007. nytimes.com/2007/02/22/garden/22depression.html?_r=1.

Collins, Nancy. 'Set Design: Something's Gotta Give'. *Architectural Digest Magazine*. 30 June 2007. architecturaldigest.com/story/somethings-gotta-give-film-sets-article.

Something's Gotta Give. Directed by Nancy Meyers. Los Angeles, CA: Columbia Pictures, 2003.

Wansink. *Slim by Design*.

31. WORKING LUNCH

Goldman, Andrew. 'Director Alejandro González Iñárritu on Leonardo DiCaprio, "Birdman" and the Importance of a Proper Lunch'. *The Wall Street Journal Magazine.* 30 November 2015. wsj.com/articles/director-alejandro-gonzalez-inarritu-on-leonardo-dicaprio-birdman-and-the-importance-of-a-proper-lunch-1448894869.

Luckerson, Victor. 'Is Lunch a Waste of Time—or a Productivity Booster?' *Time.* 16 July 2012. business.time.com/2012/07/16/the-lunch-hour-necessity-or-nuisance.

Pinsel, E. Melvin, and Ligita Dienhart. *Power Lunching: how you can profit from more effective business lunch strategy.* Chicago: Turnbull & Willoughby, 1984.

Röttgers, Kurt. *Kritik der kulinarischen Vernunft: Ein Menü der Sinne nach Kant* (*A Critique of Culinary Rationality: a menu of the senses à la Kant*). Bielefeld: transcript Verlag, 2009.

Thøgersen-Ntoumani, Cecilie, Elizabeth A. Loughren, Florence-Emilie Kinnafick, Ian Mark Taylor, Joan L. Duda, and Kenneth R. Fox. 'Changes in Work Affect in Response to Lunchtime Walking in Previously Physically Inactive Employees: a randomized trial'. *Scandinavian Journal of Medicine and Science in Sports* 25, no. 6 (February 2015): 778–87.

Weber, Daniel. ' "Lunch ist nur noch Pflicht": Interview mit Philippe Stern' (' "Lunch Is Merely a Duty": an interview with Philippe Stern'). *Neue Zürcher Zeitung Folio.* June 2006. folio.nzz.ch/2006/juni/lunch-ist-nur-noch-pflicht.

32. FAST MANNERS

Leimgruber, Walter. 'Adieu Zmittag' ('Adieu Lunch'). *Neue Zürcher Zeitung Folio.* June 2006. folio.nzz.ch/2006/juni/adieu-zmittag.

Rubin, Lawrence C., ed. *Food for Thought: essays on eating and culture*. Jefferson, NC: McFarland, 2008.

33. STRESS-FREE SLURPING

Business Wire. '"20 Worst Drinks in America" Revealed: authors of the bestselling book series "Eat This, Not That!" unveil definitive list of unhealthiest drinks in new book, "Drink This, Not That!"' 27 May 2010. businesswire.com/news/home/20100527006737/en/%E2%80%9C20-Worst-Drinks-America%E2%80%9D-Revealed.

Sandow, Erika. 'On the Road: social aspects of commuting long distances to work'. PhD diss., Umeå University, 2011. urn.kb.se/resolve?urn=urn:nbn:se:umu:diva-43674.

'Understanding the Job'. YouTube video. 4:55. Presentation by Clayton Christensen of Harvard Business School. Posted by 'University of Phoenix', 13 June 2012. youtube.com/watch?v=f84LymEs67Y.

34. THE COMFORT-FOOD EFFECT

Fisher, Helen E., Lucy L. Brown, Arthur Aron, Greg Strong, and Debra Mashek. 'Reward, Addiction, and Emotion Regulation Systems Associated with Rejection in Love'. *Journal of Neurophysiology* 104, no. 1 (July 2010): 51–60.

Hoffman, Jan. 'The Myth of Comfort Food'. *The New York Times*. 15 December 2014. well.blogs.nytimes.com/2014/12/15/the-myth-of-comfort-food.

Platte, Petra, Cornelia Herbert, Paul Pauli, and Paul A. S. Breslin. 'Oral Perceptions of Fat and Taste Stimuli Are Modulated by Affect and Mood Induction'. *PLOS ONE* 8, no. 6 (June 2013): e65006.

Troisi, Jordan D., and Shira Gabriel. 'Chicken Soup Really Is Good for the Soul: "comfort food" fulfills the need to belong'. *Psychological Science* 22, no. 6 (May 2011): 747–53.

Troisi, Jordan D., Shira Gabriel, Jaye L. Derrick, and Alyssa Geisler. 'Threatened Belonging and Preference for Comfort Food Among the Securely Attached'. *Appetite* 90, no. 1 (July 2015): 58–64.

Wagner, Heather Scherschel, Britt Ahlstrom, Joseph P. Redden, Z.M. Vickers, and Traci Mann. 'The Myth of Comfort Food'. *Health Psychology* 33, no. 12 (August 2014): 1552–57.

Xu, Xiaomeng, Lucy L. Brown, Arthur Aron, Cao Guikang, Tingyong Feng, Bianca P. Acevedo, and Xuchu Weng. 'Regional Brain Activity During Early-Stage Intense Romantic Love Predicted Relationship Outcomes After 40 Months: an fMRI assessment'. *Neuroscience Letters* 526, no. 1 (September 2012): 33–38.

35. SUPERTASTERS

Bartoshuk, Linda M. 'The Biological Basis of Food Perception and Acceptance'. *Neuroscience* 4, no. 1–2 (December 1993): 21–32.

Behrens, Maik, Howard C. Gunn, Purita C. M. Ramos, Wolfgang Meyerhof, and Stephen Wooding. 'Genetic, Functional, and Phenotypic Diversity in TAS2R38-Mediated Bitter Taste Perception'. *Chemical Senses* 38, no. 6 (April 2013): 475–84.

Benton, David. 'Role of Parents in the Determination of the Food Preferences of Children and the Development of Obesity'. *International Journal of Obesity* 28, no. 7 (July 2004): 858–69.

Breslin, Paul A. S. 'An Evolutionary Perspective on Food and Human Taste'. *Current Biology* 23, no. 9 (May 2013): R409–18.

Campbell, Michael C., Alessia Ranciaro, Alain Froment, Jibril Hirbo, Sabah A. Omar, Jean-Marie Bodo, Thomas B. Nyambo, et al. 'Evolution of Functionally Diverse Alleles Associated with PTC Bitter Taste Sensitivity in Africa'. *Molecular Biology and Evolution* 29, no. 4 (November 2011): 1141–53.

Campbell, Michael C., and Sarah A. Tishkoff. 'The Evolution of Human Genetic and Phenotypic Variation in Africa'. *Current Biology* 20, no. 4 (February 2010): R166–73.

Connor, Steve. 'Supertasters Live in a Neon-Lit World of Food Flavours'. *Independent*. 17 February 2003. independent .co.uk/news/science/supertasters-live-in-a-neon-lit-world-of-food-flavours-119278.html.

Davis, Heather A. 'Genetics Study: Africans Have Keener Sensitivity to Bitter Tastes'. *Penn Current*. 19 February 2009. penncurrent.upenn.edu/node/2378.

Fox, Arthur L. 'The Relationship Between Chemical Constitution and Taste'. *Proceedings of the National Academy of Sciences USA* 18, no. 1 (January 1932): 115–20.

Hayes, John E., Emma Louise Feeney, Alissa Nolden, and John E. Mcgeary. 'Quinine Bitterness and Grapefruit Liking Associate with Allelic Variants in TAS2R3'. *Chemical Senses* 40, no. 6 (May 2015): 437–43.

36. SOME LIKE IT HOT

Albrecht, Harro. *Schmerz: Eine Befreiungsgeschichte* (*Pain: a story of liberation*). Munich: Pattloch, 2015.

Byrnes, Nadia K., and John E. Hayes. 'Gender Differences in the Influence of Personality Traits on Spicy Food Liking and Intake'. *Food Quality and Preference* 42 (June 2015): 12–19.

Byrnes, Nadia K., and John E. Hayes. 'Personality Factors Predict Spicy Food Liking and Intake'. *Food Quality and Preference* 28, no. 1 (April 2013): 213–21.

Caterina, Michael J., Mark A. Schumacher, Makoto Tominaga, Tobias A. Rosen, Jon D. Levine, and David Julius. 'The Capsaicin Receptor: a heat-activated ion channel in the pain pathway'. *Nature* 389 (October 1997): 816–24.

Lemke, Harald. 'Der Mensch ist, was er isst: Ludwig Feuerbach als Vordenker der Gastrosophie' ('We Are What We Eat: Ludwig Feuerbach, a pioneer of gastrosophy'). *Epikur Journal für Gastrosophie*. January 2011.

Meier, Brian P., Sara K. Moeller, Miles Riemer-Peltz, and Michael D. Robinson. 'Sweet Taste Preferences and Experiences Predict Prosocial Inferences, Personalities, and Behaviors'. *Journal of Personality and Social Psychology* 102, no. 1 (January 2012): 163–74.

Rozin, Paul, and Deborah Schiller. 'The Nature and Acquisition of a Preference for Chili Pepper by Humans'. *Motivation and Emotion* 4, no. 1 (March 1980): 77–101.

37. CONDITIONS OF TASTE

Cuda-Kroen, Gretchen. 'Baby's Palate and Food Memories Shaped Before Birth'. *Morning Edition*. NPR. 8 August 2011. npr.org/2011/08/08/139033757/babys-palate-and-food-memories-shaped-before-birth.

Hartmann, Andreas. *Zungenglück und Gaumenqualen: Geschmackserinnerungen* (*Of Happy Tongues and Tormented Palates: memories of taste*). Munich: C. H. Beck, 1994.

Mennella, Julie A., Coren P. Jagnow, and Gary K. Beauchamp. 'Prenatal and Postnatal Flavor Learning by Human Infants'. *Pediatrics* 107, no. 6 (June 2001): E88.

Nooteboom, Cees. *Rituale* (*Rituals*). Frankfurt: Suhrkamp Verlag, 1985.

Proust, Marcel. *Swann's Way: in search of lost time*. Vol. 1. Translated by Lydia Davis. New York: Penguin Classics, 2004.

Rosenblum, Lawrence D. *See What I'm Saying: the extraordinary powers of our five senses*. New York: W. W. Norton & Company, 2011.

Spieler, Marlena. 'When a Food Writer Can't Taste'. *The New York Times*. 11 January 2014. nytimes.com/2014/01/12/opinion/sunday/when-a-food-writer-cant-taste.html.

38. MIND OVER MEAT

Bastian, Brock, Steve Loughnan, Nick Haslam, and Helena R. M. Radke. 'Don't Mind Meat? The Denial of Mind to Animals Used for Human Consumption'. *Personality and Social Psychology Bulletin* 38, no. 2 (October 2011): 247–56.

Foer, Jonathan Safran. *Eating Animals*. New York: Little, Brown and Company, 2009.

Joy, Melanie. *Why We Love Dogs, Eat Pigs, and Wear Cows: an introduction to carnism*. San Francisco: Red Wheel / Weiser, 2010.

Loughnan, Steve, Bastian Brock, and Nick Haslam. 'The Psychology of Eating Animals'. *Current Directions in Psychological Science* 23, no. 2 (April 2014): 104–108.

White, Gilbert. *The Natural History and Antiquities of Selborne*. London: Bensley for B. White and Son, 1789.

39. HOORAY FOR HAPTICS

Baumgarthuber, Christine. 'Red Holidays of Genius'. *New Inquiry* (blog). 4 February 2014. thenewinquiry.com/blog/red-holidays-of-genius.

Emmerich, A. 'Der Kochkünstler: Ein Gespräch mit Adrià Ferran' ('The Artistic Chef: an interview with Adrià Ferran'). *Zeit* 1 (2007).

Gebhardt, Ulrike. '"Der Tastsinn ist ein Lebensprinzip": Interview mit Martin Grunwald, Leiter des Haptik-Forschungslabors der Universität Leipzig' ('"The Sense of Touch Is a Life Principle": Interview with Martin Grunwald, Director at the Research Laboratory for Haptics at the University of Leipzig'). *Spektrum*. 25 July 2014. spektrum.de/news/ohne-tastsinn-gibt-es-kein-leben/1302125.

Marinetti, Filippo Tommaso. *The Futurist Cookbook*. Translated by Suzanne Brill. London: Penguin Modern Classics, 2014.

Phil. 'How to Make Heston Blumenthal Fat Duck Style Hot and Iced Waitrose Mulled Cider or Mulled Wine Recipe'. *In Search of Heston*. 16 December 2003. insearchofheston.com/2013/12/how-to-make-heston-blumenthal-fat-duck-style-hot-and-iced-waitrose-mulled-cider-or-mulled-wine-recipe.

Stroh, S. 'Haptische Wahrnehmung und Textureigenschaften von Lebensmitteln' ('Haptic Perception and the Structural Properties of Food Items'). In *Der bewegte Sinn* (*The Animated Sense*) edited by Matin Grunwald and Lothar Beyer, 195–97. Basel: Birkhäuser Verlag, 2013.

Zuber, Helene. 'Im Mund explodiert' ('Exploded in the Mouth'). *Der Spiegel* 52. 25 December 2000. spiegel.de/spiegel/print/d-18124620.html.

40. TASTE THE AROMA

'09 Dunkin' Donuts Flavor Radio'. YouTube video. 2:33. Posted by 'CheilGlobal', 26 June 2012. youtube.com/watch?v=aHg0xFZQzYI#t=39.

Aiello, Leslie, and Christopher Dean. *An Introduction to Human Evolutionary Anatomy*. New York: Academic Press, 1990.

Bensafi, M., E. Iannilli, J. Gerber, and T. Hummel. 'Neural Coding of Stimulus Concentration in the Human Olfactory and Intranasal Trigeminal Systems'. *Neuroscience* 154, no. 2 (June 2008): 832–38.

Brillat-Savarin. *The Physiology of Taste.*

Bushdid, C., M. O. Magnasco, L. B. Vosshall, and A. Keller. 'Humans Can Discriminate More Than 1 Trillion Olfactory Stimuli'. *Science* 343, no. 6177 (March 2014): 1370–72.

Croy, Ilona, Steven Nordin, and Thomas Hummel. 'Olfactory Disorders and Quality of Life: an updated review'. *Chemical Senses* 39, no. 3 (January 2014): 185–94.

Escapist. 'Poll: If You Had to Lose One Sense, What Would It Be?'escapistmagazine.com/forums/read/18.273130-Poll-If-you-had-to-lose-one-sense-what-would-it-be?.

Hatt, Hanns, and Regine Dee. *Das kleine Buch vom Riechen und Schmecken* (*The Little Book of Smell and Taste*). Munich: Albrecht Knaus Verlag, 2012.

Hirsch, Alan R. 'Effects of Ambient Odors on Slot-Machine Usage in a Las Vegas Casino'. *Psychology and Marketing* 12, no.7 (October 1995): 585–94.

Hummel, Thomas, Karo Rissom, Jens Reden, Aantje Hähner, Mark Weidenbecher, and Karl-Bernd Hüttenbrink. 'Effects of Olfactory Training in Patients with Olfactory Loss'. *Laryngoscope* 119, no. 3 (March 2009): 496–99.

Kobal, G., Thomas Hummel, B. Sekinger, S. Barz, Stephan Roscher, and S. R. Wolf. '"Sniffin' Sticks": screening of olfactory performance'. *Rhinology* 34, no. 4 (December 1996): 222–26.

Kollndorfer, Kathrin, Florian Ph. S. Fischmeister, Ksenia Kowalczyk, Elisabeth Hoche, Christian A. Mueller, Siegfried Trattnig, and Veronika Schöpf. 'Olfactory Training Induces

Changes in Regional Functional Connectivity in Patients with Long-Term Smell Loss'. *NeuroImage: Clinical* 9 (2015): 401–10.

Lemke, Harald. *Die Kunst des Essens: Eine Ästhetik des kulinarischen Geschmacks* (*The Art of Eating: the aesthetics of culinary taste*). Bielefeld: transcript Verlag, 2007.

Neville, Kevin R., and Lewis B. Haberly. 'Olfactory Cortex'. In *The Synaptic Organization of the Brain*, 5th ed., edited by Gordon M. Shepherd, 415–54. Oxford: Oxford University Press, 2004.

Shepherd, Gordon M. 'The Human Sense of Smell: are we better than we think?' *PLOS* 2, no. 5 (May 2004): e146. doi: 10.1371/journal.pbio.0020146.

Shepherd. *Neurogastronomy.*

Wysocki, Charles J., Kathleen M. Dorries, and Gary K. Beauchamp. 'Ability to Perceive Androstenone Can Be Acquired by Ostensibly Anosmic People'. *Proceedings of the National Academy of Sciences of the United States of America* 86, no. 20 (October 1989): 7976–78.

41. CHEW ON THIS

Elias, Norbert. *The Civilizing Process: sociogenetic and psychogenetic investigations.* Translated by Edmund Jephcott. Oxford: Blackwell, 2000.

'Food Hacking: electric fork'. YouTube video. 15:33. Posted by 'Munchies', 18 January 2016, youtube.com/watch?v=95rrDcdctlE.

Wilson, Bee. *Consider the Fork: a history of how we cook and eat.* New York: Basic Books, 2012.

42. THE PINEAPPLE FALLACY

Benton, David. 'Role of Parents in the Determination of the Food Preferences of Children and the Development of Obesity'. *International Journal of Obesity* 28, no.7 (July 2004): 858–69.

Felser, Georg. *Werbe- und Konsumentenpsychologie* (*Advertising and Consumer Psychology*). Berlin, Heidelberg: Springer, 2015.

Reitmeier, Simon. *Warum wir mögen, was wir essen: Eine Studie zur Sozialisation der Ernährung* (*Why We Like What We Eat: a study of the socialisation of food*). Bielefeld: transcript Verlag, 2013.

Shepherd, R., and M. Raats, eds. *The Psychology of Food Choice*. Wallingford, CT: Cabi Publishing, 2010.

Von Engelhardt, Dietrich. *Geschmackskulturen: Vom Dialog der Sinne beim Essen und Trinken* (*Taste Cultures: how our senses communicate when we eat and drink*). Edited by Rainer Wild Frankfurt: Campus Verlag, 2005.